Advanced Bioactive Inorganic Materials for Bone Regeneration and Drug Delivery

Advanced Bioactive Inorganic Materials for Bone Regeneration and Drug Delivery

Edited by
Chengtie Wu ▪ Jiang Chang ▪ Yin Xiao

CRC Press
Taylor & Francis Group
Boca Raton London New York

CRC Press is an imprint of the
Taylor & Francis Group, an **informa** business

CRC Press
Taylor & Francis Group
6000 Broken Sound Parkway NW, Suite 300
Boca Raton, FL 33487-2742

First issued in paperback 2019

© 2013 by Taylor & Francis Group, LLC
CRC Press is an imprint of Taylor & Francis Group, an Informa business

No claim to original U.S. Government works

ISBN-13: 978-1-4665-5199-2 (hbk)
ISBN-13: 978-0-367-38024-3 (pbk)

Library of Congress Cataloging-in-Publication Data
Advanced bioactive inorganic materials for bone regeneration and drug delivery / editors, Chengtie Wu, Jiang Chang, Yin Xiao. p. ; cm. Includes bibliographical references and index. ISBN 978-1-4665-5199-2 (hardcover : alk. paper) I. Wu, Chengtie. II. Chang, Jiang (Professor) III. Xiao, Yin. [DNLM: 1. Bone Regeneration--drug effects. 2. Biocompatible Materials--therapeutic use. 3. Tissue Engineering--methods. WE 200]

617.4'71061--dc23   2012050711

**Visit the Taylor & Francis Web site at
http://www.taylorandfrancis.com**

**and the CRC Press Web site at
http://www.crcpress.com**

Contents

Preface

Bioceramics play an important role in repairing and regenerating bone defects. Annually, more than 500,000 bone graft procedures are performed in the United States and approximately 2.2 million are conducted worldwide. The estimated cost of these procedures approaches $2.5 billion per year. Around 60% of the bone graft substitutes available on the market involve bioceramics. It is reported that bioceramics in the world market increase by 9% per year. For this reason, the research of bioceramics has been one of the most active areas during the past several years. Considering the significant importance of bioceramics, our goal was to compile this book to review the latest research advances in the field of bioceramics. The text also summarizes our work during the past 10 years in an effort to share innovative concepts, design of bioceramics, and methods for material synthesis and drug delivery. We anticipate that this text will provide some useful information and guidance in the bioceramics field for biomedical engineering researchers and material scientists.

Information on novel mesoporous bioactive glasses and silicate-based ceramics for bone regeneration and drug delivery are presented. Mesoporous bioactive glasses have shown multifunctional characteristics of bone regeneration and drug delivery due to their special mesopore structures, whereas silicate-based bioceramics, as typical third-generation biomaterials, possess significant osteostimulation properties. Silica nanospheres with a core-shell structure and specific properties for controllable drug delivery have been carefully reviewed—a variety of advanced synthetic strategies have been developed to construct functional mesoporous silica nanoparticles with a core-shell structure, including hollow, magnetic, or luminescent, and other multifunctional core-shell mesoporous silica nanoparticles. In addition, multifunctional drug delivery systems based on these nanoparticles have been designed and optimized to deliver the drugs into the targeted organs or cells, with a controllable release fashioned by virtue of various internal and external triggers. The novel 3D-printing technique to prepare advanced bioceramic scaffolds for bone tissue engineering applications has been highlighted, including the preparation, mechanical strength, and biological properties of 3D-printed porous scaffolds of calcium phosphate cement and silicate bioceramics. Three-dimensional printing techniques offer improved large-pore structure and mechanical strength. In addition, biomimetic preparation and controllable crystal growth as well as biomineralization of bioceramics are summarized, showing the latest research progress in this area. Finally, inorganic and organic composite materials are reviewed for bone regeneration and gene delivery. Bioactive inorganic and organic composite materials offer unique biological, electrical, and mechanical properties for designing excellent bone regeneration or gene delivery systems.

It is our sincere hope that this book will update the reader as to the research progress of bioceramics and their applications in bone repair and regeneration. It will be the best reward to all the contributors of this book if their efforts herein in some way help readers in any part of their study, research, and career development.

Acknowledgments

The authors thank the Pujiang Talent Program of Shanghai (12PJ1409500), Recruitment Program of Global Young Talent, China (Chengtie Wu), Natural Science Foundation of China (Grants 81201202 and 81190132), Shanghai Municipal Natural Science Foundation (12ZR1435300), and ARC Discovery (DP120103697) for the support of research published in this book.

Editors

Chengtie Wu, Ph.D., works at the Shanghai Institute of Ceramics, Chinese Academy of Sciences (SIC, CAS). He completed his Ph.D. in 2006, and then worked at the University of Sydney, Australia; Dresden University of Technology, Germany; and Queensland University of Technology as Vice Chancellor Research Fellow and Alexander von Humboldt Fellow. In 2012, Dr. Wu was recruited to work in SIC, CAS, with the One-Hundred Talent Program of Chinese Academy of Sciences and One-Thousand Young Talent Program of China. Dr. Wu's research focuses on advanced bioactive inorganic materials for bone tissue engineering and drug delivery application. He has published more than 60 peer-reviewed journal papers, including 11 papers in *Biomaterials*, the top journal in the field of biomaterials. The published papers have been cited more than 700 times. Dr. Wu has been awarded five patents; two of the technology patents have been transferred to companies.

Jiang Chang, Ph.D., received a Ph.D. in 1991 in chemistry from the Technical University of Darmstadt in Germany. He then spent 2 years as a postdoctoral research fellow at the Medical University of Luebeck in Germany, where he worked on collagen-based biomaterials and artificial skin for wound healing. This was followed by a 4-year (1993–1997) appointment as a research fellow at the School of Medicine, University of Auckland, New Zealand, where he conducted research on cell–biomaterials interactions and cartilage repair. In 1997, Dr. Chang moved to the United States and was appointed as a research assistant professor in the medical school at New York University from 1997 to 1999, and as a research scientist at Johnson & Johnson from 1999 to 2000. There he studied biomaterials for bone and cartilage regeneration. Since 2001, Dr. Chang has been the director of the Biomaterials and Tissue Engineering Research Center at the Shanghai Institute of Ceramics, Chinese Academy of Sciences. His research focuses on bioactive materials for tissue regeneration and tissue engineering. He is also adjunct professor of China-East Normal University (since 2006) and Shanghai Jiaotong University (since 2008). Dr. Chang is an editorial board member of several scientific journals including *Ceramics International, Bio-Medical Materials and Engineering*, and the *Journal of Inorganic Materials*. In 2011, he was elected as vice president of the Interdisciplinary Research Society for Bone and Joint Injectable Biomaterials. Dr. Chang has over 200 scientific papers published in international peer-reviewed scientific journals including *NanoLetters, Advanced Materials, European Cells and Materials*, the *Journal of Controlled Release*, and *Biomaterials*; and 45 patents in the field of biomedical materials. Due to his contribution in the field of biomaterials, he was awarded a Fellow of International Union of Societies for Biomaterials Science and Engineering (FBSE) in 2012.

Yin Xiao, Ph.D., is a biomedical scientist, dentist, and professor of medical engineering at Queensland University of Technology. He completed his undergraduate and postgraduate education at Wuhan University (1986 and 1991) and University of Queensland (Ph.D., 2000). He has been awarded several fellowships and was promoted to associate professor in 2005 and professor in 2012. He is the group leader of the bone and clinical research program. Dr. Xiao has an international standing in the research areas of biomaterials, bone biology, tissue engineering, regenerative medicine, and orthopedic and dental research, which is reflected by his numerous publications in highly regarded international journals and invited presentations as a keynote speaker at international conferences. He also serves as an invited panel member to assess higher-degree training programs for several universities, is a reviewer for a number of international and national grant bodies, an editorial board member on a series of scientific journals, and an invited reviewer of over 40 journals. Dr. Xiao is also an adjunct professor at Wuhan University, Sun Yat-sen University, Fujian Medical University, and Griffith University.

Contributors

Ahmed Ballo
Department of Oral Health Sciences
Faculty of Dentistry
University of British Columbia
Vancouver, BC, Canada
and
Dental Implant and Osseointegration
    Research Chair
College of Dentistry
King Saud University
Riyadh, Saudi Arabia

Michael Gelinsky
Centre for Translational Bone, Joint,
    and Soft Tissue Research
University Hospital Dresden and
    Faculty of Medicine Carl Gustav
    Carus
Dresden University of Technology
Dresden, Germany

Zhongru Gou
Zhejiang–California International
    NanoSystems Institute
Zhejiang University
Hangzhou, China

Kaili Lin
Shanghai Institute of Ceramics
Chinese Academy of Sciences
Shanghai, China

Anja Lode
Centre for Translational Bone, Joint,
    and Soft Tissue Research
University Hospital Dresden and
    Faculty of Medicine Carl Gustav
    Carus
Dresden University of Technology
Dresden, Germany

Yongxiang Luo
Centre for Translational Bone, Joint,
    and Soft Tissue Research
University Hospital Dresden and
    Faculty of Medicine Carl Gustav
    Carus
Dresden University of Technology
Dresden, Germany

Wei Xia
BIOMATCELL
Gothenburg, Sweden
and
Department of Engineering Science
Uppsala University
Uppsala, Sweden

Yufeng Zhang
The State Key Laboratory
    Breeding Base of Basic Science
    of Stomatology (Hubei-MOST)
    and Key Laboratory of Oral
    Biomedicine Ministry of
    Education
School and Hospital of Stomatology
Wuhan University
Wuhan, People's Republic of China

Yufang Zhu
School of Materials Science and
    Engineering
University of Shanghai for Science
    and Technology
Shanghai, People's Republic of
    China

1

Mesoporous Bioactive Glasses for Drug Delivery and Bone Tissue Regeneration

Chengtie Wu, Jiang Chang, and Yin Xiao

CONTENTS

1.1 Introduction

1.1.1 Conventional Bioactive Glasses

Bioactive glasses have played an increasingly important role in bone tissue regeneration application by virtue of their generally excellent osteoconductivity, osteostimulation, and degradation rate (Chen et al. 2006; Jones et al. 2006, 2007; Hench and Thompson 2010; Misra et al. 2010; Wu, Hill, et al. 2011). The melt-derived bioactive glass, called 45S5® bioglass, was pioneered by Hench (1991, 1998) and was first developed using traditional melt methods at high temperature (1300°C–1500°C). The 45S5 bioglass has been regarded as bioactive bone regeneration materials that are able to bond closely with the host bone tissue (Hench 1991). The mechanism behind new bone formation on bioactive glasses is closely associated with the release of Na^+ and Ca^{2+} ions and the deposition of a carbonated hydroxyapatite (CHAp) layer. The apatite layer forms a strong chemical bond between 45S5 bioglass and the host bone (Hench 1991). Further studies have also shown that the Ca and Si released from the 45S5 contribute to its bioactivity, as both Ca and Si are found to stimulate osteoblast proliferation and differentiation (Xynos et al. 2000; Gough, Jones, et al. 2004; Gough, Notingher, et al. 2004; Jones et al. 2006, 2007; Hoppe et al. 2011). Xynos et al. (2000) further found that 45S5 bioglass is able to enhance the expression of a potent osteoblast mitogenic growth factor, insulin-like growth factor II (IGF-II) (Xynos et al. 2000; Wu, Chang, et al. 2011). The 45S5 bioglass is still considered to be the gold standard for bioactive glasses, although melt-derived bioactive glasses have a number of limitations (Wu, Chang, et al. 2011). One of these limitations is the fact that it needs to be melted at a very high temperature (>1300°C) and another is its lack of microporous structure inside the materials with low specific surface area; therefore, the bioactivity of melt-derived bioactive glasses will mainly depend on the contents of SiO_2 (Wu, Chang, et al. 2011). Generally, the bioactivity of melt-derived bioactive glasses will decrease with the increase of SiO_2 contents (Hench 1998; Hench and Polak 2002; Hench and Thompson 2010). When the SiO_2 content exceeds 60%, the bioactive glass is not able to induce the formation of CHAp layers even after several weeks in simulated body fluid (SBF) and it fails to bond to either bone or soft tissues (Arcos and Vallet-Regi 2010). The main reason is that high SiO_2-containing glasses prepared by the melt-derived method have a stable net structure and do not easily release Na^+ and Ca^{2+}, leading to insufficient OH^- groups on the surface of glasses to induce apatite formation. In the early 1990s, in an effort to overcome the limitation of melt-derived bioactive glasses, Li et al. (1991) prepared sol-gel–derived bioactive glasses. They demonstrated that this class of materials was bioactive in a wider compositional range when compared with traditional melt-derived bioactive glasses. The glasses in the SiO_2-CaO-P_2O_5 system had a silica content of up to 90% and were still capable of inducing the formation of apatite layers, compared to the 60% SiO_2 boundary of the melt-derived bioactive glasses (Cerruti and Sahai 2006). Due to its larger surface area and porosity properties derived from the sol-gel process, the range of bioactive compositions for sol-gel–derived bioglasses is wider, and as compared to melt-derived bioactive glasses, these bioactive glasses exhibit higher bone bonding rates coupled with excellent degradation and resorption properties (Zhong and Greenspan 2000; Hamadouche et al. 2001; Arcos and Vallet-Regi 2010). Although sol-gel–derived bioactive glasses have better composition range and bioactivity than melt-derived bioactive glasses, the micropore distribution is not uniform and is inadequate for efficient drug loading and release (Arcos et al. 2009; Wu, Chang, et al. 2011; Wu and Chang 2012).

1.1.2 Mesoporous SiO₂ and Mesoporous Bioactive Glass (MBG) Materials

In the past 20 years, mesoporous materials have attracted great attention due to their significant feature of large surface area, ordered mesoporous structure, tunable pore size and volume, and well-defined surface property. They have many potential applications, such as catalysis, adsorption/separation, and biomedicine (Zhao, Feng, et al. 1998). The studies of mesoporous materials have been expanded into the field of biomaterials science for drug delivery and bone regeneration application. It has been systematically investigated that the *in vitro* apatite formation of different types of mesoporous materials demonstrated that an apatite-like layer can be formed on the surfaces of Mobil Composition of Matters (MCM)-48, hexagonal mesoporous silica (SBA-15), phosphorous-doped MCM-41, and bioglass-containing MCM-41, allowing their use in biomedical engineering for tissue regeneration (Lopez-Noriega et al. 2006; Vallet-Regi 2006; Vallet-Regi, Ruiz-Gonzalez, et al. 2006). Mesoporous silica (SiO₂) was also used for the study of efficient drug delivery. It was found that mesoporous silica is an attractive material due to its good biocompatibility, low cytotoxicity, tailored surface charge, and enormous possibilities for organic functionalization (Vallet-Regi, Ruiz-Gonzalez, et al. 2006; Vallet-Regi et al. 2007; Manzanoab and Vallet-Regi 2010). In addition, mesoporous silica present unique mesopore structures and porosities with large surface area, high pore volume, and narrow mesopore channels that allow the adsorption of drugs and biomolecules into their well-ordered pores and cavities to be then locally released (Vallet-Regi et al. 2007). However, although pure mesopore silica has shown to be an excellent drug delivery system, it has generally too low activity of *in vitro* apatite mineralization to be considered as bioactive bone grafts (Horcajada et al. 2004; Izquierdo-Barba et al. 2005). In addition, mesoporous SiO₂ does not show obvious degradation, which limits their application for bone regeneration (Wu and Chang 2012).

In bone reconstruction surgeries, osteomyelitis caused by bacteria infection is one of the main complications. Conventional treatments include systemic antibiotic administration, surgical debridement, wound drainage, and implant removal (Zhao, Yan, et al. 2008). These approaches, however, are not always efficient and the patients may suffer from extra surgeries. A new method to solve this problem is to introduce a local drug release system into the implant site (Mourino and Boccaccini 2010). The advantages of this treatment include high delivery efficiency, continuous action, reduced toxicity, and convenience to the patients (Zhao, Yan, et al. 2008; Zhu, Zhang, Wu, et al. 2011). Therefore, to overcome the limitations of conventional bioactive glasses (without well-ordered mesopore structures for drug delivery) and pure mesopore SiO₂ (low bioactivity), it is of great importance to design and develop a new class of biomaterials that combine efficient drug delivery and excellent bioactivity. Yan et al. (2004, 2006) for the first time prepared mesoporous bioactive glasses (MBG) in 2004 by combining the sol-gel method and supramolecular chemistry of surfactants. Their study has opened a new direction for applying nanotechniques to regenerative medicine by coupling drug delivery with bioactive materials. These materials are based on a CaO-SiO$_2$-P$_2$O$_5$ composition and have a highly ordered mesopore channel structure with a pore size ranging from 5 to 20 nm. Compared to conventional nonmesoporous bioactive glass (NBG), the MBG possesses a more optimal surface area, pore volume, ability to induce *in vitro* apatite mineralization in simulated body fluids, and excellent cytocompatibility (Yan et al. 2006; Leonova et al. 2008; Garcia et al. 2009; Alcaide et al. 2010). The study of MBG for drug delivery and bone tissue engineering has been becoming a hot area of research during the past 5 years (Wu and Chang 2012). In this chapter, we will highlight the recent advances of MBG materials for drug and growth factor delivery and bone regeneration applications.

1.2 Preparation, Compositions, and Main Forms of MBG

1.2.1 Preparation Methods and Compositions of MBG

The preparation method of MBG is similar with that for mesoporous SiO$_2$, in which the supramolecular chemistry has been incorporated into the sol-gel process. In this strategy, the incorporation of structure-directing agents (e.g., CTAB, P123, and F127) is essential for obtaining well-ordered structures. Under appropriate synthesis conditions, these molecules self-organize into micelles. Micelles link the hydrolyzed silica precursors through the hydrophilic component and self-assemble to form an ordered mesophase (Arcos et al. 2009; Arcos and Vallet-Regi 2010). Then the reaction mixture of bioactive glasses and structure-directing agents underwent an evaporation-induced self-assembly (EISA) process. A general definition of EISA is the spontaneous organization of materials through noncovalent interactions (hydrogen bonding, van der Waals forces, electrostatic forces, etc.) with no external intervention (Brinker et al. 1999). In the EISA process of MBG, the surfactants assemble into micelles, spherical or cylindrical structures that maintain the hydrophilic parts of the surfactants in contact with the composition of bioactive glasses (Si, Ca, and P elements, etc.) while shielding the hydrophobic parts within the micellar interior. Once the mixture is dried and the surfactant removed, a well-ordered mesoporous structure will be obtained, exhibiting high surface area and porosity (Arcos and Vallet-Regi 2010; Wu and Chang 2012).

As pure mesoporous SiO$_2$ lacks adequate bioactivity, MBG has been developed with multicomponents based on SiO$_2$-CaO or SiO$_2$-CaO-M$_x$O$_y$ (M: P, Mg, Zn, or/and Fe, etc.) to enhance their bioactivity and special functions.

Generally, MBG with high contents of SiO_2 has a more ordered mesopore structure, and higher specific surface area and pore volume than those of low-SiO_2 MBG. The incorporation of Ca or P into a mesoporous SiO_2 system significantly decreased its surface area and pore volume. Similarly, the incorporation of divalent ions (Mg, Zn, Cu, or Sr), trivalent ions (Ce, Ga, or B), and tetravalent ions (Zr) into a SiO_2-CaO-P_2O_5 MBG system also significantly decreased its mesoporous properties (surface area and pore volume) (see Table 1.1). It is interesting that the composition of MBG seems to have no significant effect on the mesopore size, which is usually in the range of 3 to 5 nm. The results indicated that the incorporation of additional ions into an MBG system may disrupt the ordered orientation of SiO_4^{4-} during the self-assembly reaction, which may result in potential structural defects in the atomic array and further change the shape and structures of mesopores (Wu, Fan, et al. 2011). However, the MBG with varied components still possesses high surface area (in the range of 150 to 500 m^2/g) and pore volume (in the range of 0.2 to 0.6 cm^3/g; see Table 1.1) (Wu and Chang 2012).

1.2.2 Different Forms of MBG: Particles, Fibers, Scaffolds, and Composites

MBG has been prepared as particles, fibers, spheres, 3D scaffolds, and composites with a well-ordered mesoporous channel structure and excellent bioactivity for drug delivery and bone regeneration application (Wu and Chang 2012). MBG particles were first synthesized in 2004 by Yan and colleagues. The size of the obtained MBG particles was around several tens of micrometers. The particles contained highly ordered mesoporous channels (5 nm) with high surface area and pore volume. After that, Lei, Chen, Wang, Zhao, Du, et al. (2009) synthesized MBG powders with high specific surface area by using acetic acid as a structure-assisted agent and hydrolysis catalyst. MBG powders with different compositions (58S and 77S) were prepared using P123 and hydrothermal treatment, both of which were shown to have excellent *in vitro* bioactivity (Xia and Chang 2006, 2008). By using the same method, Li, Wang, He, et al. (2008) prepared Mg-, Zn-, or Cu-containing multicomponent MBG particles. Wu, Wei, et al. (2010) synthesized CaO-SiO_2 mesoporous MBG particles for hemostatic application. Recently, our group developed a simple method to prepare MBG particles without hydrothermal treatment, which is suitable for large-mass production of MBG particles (Feng and Chang 2011). Nanosized mesoporous MBG particles with a size of around 100 nm were prepared and could be used as bioactive fillers to infiltrate into dentinal tubules, carry/release antibiotics, and induce *in vitro* mineralization (see Figure 1.1).

Hong, Chen, Jing, Fan, Guo, et al. (2010) prepared ultrathin MBG fibers by electron spin techniques. In their study, ultrathin MBG fibers, with hierarchical nanoporosities and high matrix homogeneities, were synthesized using electrospinning techniques and P123-PEO cotemplates. At the same

TABLE 1.1

Effect of Composition on the Characteristics of Mesopore Structures

MBG with Different Compositions	Surface Area (m²/g)	Pore Volume (cm³/g)	Pore Size (nm)	References
100Si	490		3.6	Wu, Wei, et al. 2010
95Si5Ca	467		3.7	
90Si10Ca	438		3.5	
100Si	310	0.356	4.2	Garcia et al. 2011
97.5SiP2.5 (TEP)	270	0.308	4.4	
97.5SiP2.5 (H_3PO_4)	152	0.235	4.8	
80Si15Ca5P	351	0.49	4.6	Yan et al. 2004
70Si15Ca5P	319	0.49	4.6	
60Si15Ca5P	310	0.43	4.3	
100Si	384	0.40	4.9	Zhu et al. 2008
90Si5Ca5P	330	0.35	4.9	
80Si15Ca5P	351	0.36	4.8	
70Si25Ca5P	303	0.33	4.8	
80Si10Ca5P5Fe	260	0.26	3.5	Li, Wang, Hua, et al. 2008
80Si5Ca5P10Fe	334	0.30	3.6	
80Si0Ca5P15Fe	367	0.36	3.7	
80Si15Ca5P	342	0.38	3.62	Li, Wang, He, et al. 2008; Zhu, Li, et al. 2011
80Si10Ca5P5Mg	274	0.35	3.31	
80Si10Ca5P5Zn	175	0.23	3.33	
80Si10Ca5P5Cu	237	0.31	3.66	
80Si10Ca5P5Sr	247	0.31	3.66	
80Si15Ca5P	515	0.58	4.7	Salinas et al. 2011
76.5Si15Ca5P3.5Ce	397	0.38	2.9	
76.5Si15Ca5P3.5Ga	335	0.31	3.8	
80Si15Ca5P	265	0.33	5.29	Wu, Miron, et al. 2011
75Si15Ca5P5B	234	0.24	5.28	
70Si15Ca5P10B	194	0.21	5.09	
80Si15Ca5P	317	0.37	4.1	Zhu, Zhang, Wu, et al. 2011
80Si10Ca5P5Zr	287	0.32	3.7	
80Si5Ca5P10Zr	278	0.33	4.1	
80Si5P15Zr	277	0.27	3.4	

Source: Wu, C., and Chang, J., 2012. Mesoporous bioactive glasses: Structure characteristics, drug/growth factor delivery and bone regeneration application. *Interface Focus* 2: 292–306.

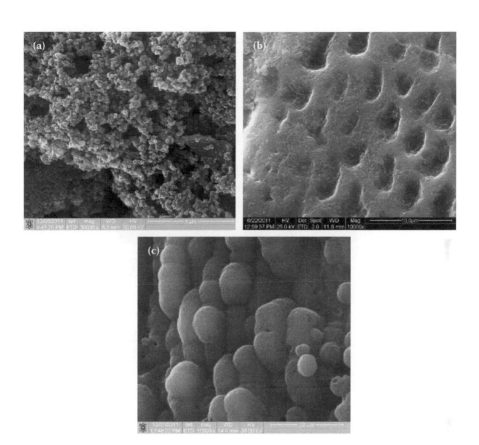

FIGURE 1.1
(a) Nanosized MBG particles, (b) the particles filling dentin tubules, and the induced apatite mineralization in SBF (c).

time, by controlling the electrospinning conditions they were able to prepare MBG fibers with hollow cores and mesoporous walls; these fibers were found to be highly bioactive for drug delivery (Hong, Chen, Jing, Fan, Gu, et al. 2010; Wu, Chang, et al. 2011; Wu and Chang 2012).

It is interesting that MBG could be prepared as uniform spheres with the size range from nanometer to millimeter. A millimeter-sized MBG sphere with a well-ordered mesopore channel structure was prepared by using the method of alginate cross-linking with Ca²⁺ ions (see Figure 1.2). The large-size MBG spheres could not only support the adhesion of bone marrow stromal cells (BMSC) but also control the delivery of proteins (Wu, Zhang, Ke, et al. 2010). Yun et al. (2009) prepared hierarchically mesoporous–macroporous MBG spheres with the size of several hundred micrometers in chloroform by the triblock copolymer templating and sol-gel technique. The spheres have well interconnected pore structures and excellent *in vitro* bioactivity. Mesoporous hollow bioactive glass microspheres with a uniform diameter

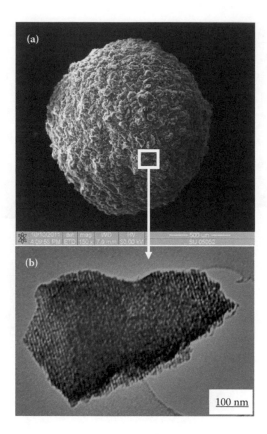

FIGURE 1.2
(a) Large-size MBG beads with a (b) well-ordered mesoporous channel structure.

range of 2 to 5 μm, and a mesoporous shell (500 nm) can be prepared by a sol-gel method using polyethylene glycol (PEG) as a template (Lei, Chen, Wang, and Zhao 2009). Zhao et al. (2010) prepared MBG microsphere with high P_2O_5 contents (up to 15%), and an approximate size of 4 to 5 μm, by using cosurfactants of P123 and cetyltrimethylammonium bromide (CTAB). Luminescent calcium silicate MBG microspheres with a diameter of 400 nm and the mesopore size of 6 nm were recently developed by Kang et al. (2011). Yun et al. (2010) produced MBG nanospheres with both a large specific surface area (1040 m^2/g) and pore volume (1.54 cm^3/g), while the size of MBG nanospheres can be further controlled over a diameter range of 20 to 200 nm by the addition of various amounts of CaO. Novel porous MBG nanospheres (80–150 nm) were prepared by a hydrothermal method, which had excellent mineralization ability (Wu and Chang 2012).

MBG can also be prepared as 3D porous scaffolds for bone tissue engineering and drug-delivery applications (Wu and Chang 2012). Currently, there are three methods to prepare MBG scaffolds. The first MBG scaffold

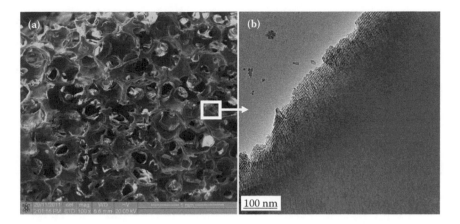

FIGURE 1.3
MBG scaffolds prepared by polyurethane templating method with (a) large pore size and (b) well-ordered mesopore channels.

was prepared by the porogen method. Yun et al. (2008) applied methyl cellulose as the porogen to prepare porous MBG scaffolds with large-pore size of 100 μm. The second is the polymer template method, which is widely used. We prepared MBG scaffolds with a large pore size of 400 μm by using polyurethane sponge as a porous template (Zhu et al. 2008). At the same time, Li et al. (2007) also prepared MBG scaffolds using the same method. After that, we have developed a series of MBG scaffolds with varying composition for drug delivery and bone tissue engineering application (Zhu, Zhang, et al. 2011; Wu, Fan, Gelinsky, et al. 2011; Wu, Fan, Zhu, et al. 2011; Wu, Miron, et al. 2011; Zhu, Zhang, et al. 2011). The prepared scaffolds possess large pores with the size of 300 to 500 μm and well-ordered mesopores with the pore size of 5 nm (see Figure 1.3). The advantages of the MBG scaffolds prepared by the polyurethane sponge template method are highly interconnective pore structure and controllable pore size (porosity), while the disadvantage is the low mechanical strength of the material (Wu, Zhang, et al. 2010). To better control the pore morphology, pore size, and porosity, a 3D plotting technique (also called direct writing or printing) has been developed to prepare porous MBG scaffolds. The significant advantage of this technique is that the architectures of the scaffolds can be concisely controlled by layer-by-layer plotting under mild conditions (Miranda et al. 2006, 2008; Franco et al. 2010). Yun et al. (2007) and Garcia et al. (2011) prepared hierarchical 3D porous MBG scaffolds using a combination of double polymer template and rapid prototyping techniques. In those studies, MBG gel was mixed with methylcellulose and then printed and sintered at 500°C to 700°C to remove polymer templates and obtain MBG scaffolds. The main limitation of this method for preparing MBG scaffolds is the need of methylcellulose and the additional sintering procedure. Although the obtained MBG scaffolds have

FIGURE 1.4
3D-printed MBG scaffolds with controllable pore architecture.

uniform pore structure, their mechanical strength is compromised because of the incorporation of methycellulose, which results in some micropores. A new facile method was recently used to prepare hierarchical and multifunctional MBG scaffolds with controllable pore architecture, excellent mechanical strength, and mineralization ability for bone regeneration application by a modified 3D-printing technique using polyvinyl alcohol (PVA) as a binder. The obtained 3D-printing MBG scaffolds possess a high mechanical strength that is about 200 times that of the MBG scaffolds prepared using traditional polyurethane foam as templates. They have highly controllable pore architecture (see Figure 1.4) and excellent apatite-mineralization ability and sustained drug-delivery property (Wu, Luo, et al. 2011; Wu and Chang 2012).

MBG, as bioactive phase, can be incorporated into polymers to improve their bioactivity and drug-delivery property. For this aim, MBG/polymer composites have been developed in the past several years (Wu and Chang 2012). Some research has prepared MBG/PLGA and MBG/polycaprolactone composite microspheres and scaffolds with improved drug-delivery ability and *in vitro* bioactivity (Li, Shi, et al. 2008; Li et al. 2009; Wei et al. 2009). MBG was used to coat the surface of macroporous poly(L-lactic acid) (PLLA) scaffolds by Zhu, Zhang, et al. (2011). MBG-coated PLLA scaffolds showed an improved bioactivity and drug-delivery property. Recently, calcium-silicate–based MBG/silk composite film with excellent osteoconductivity has been prepared (Zhu, Wu, et al. 2011). Xia et al. (2008) have prepared a dual-drug delivery system based on MBG/polypeptide graft copolymer nanomicelle composites. MBG powders can be incorporated into alginate microspheres and PLGA films to control drug delivery and improve bioactivity (Wu, Ramaswamy, et al. 2009; Wu, Zhu, et al. 2010). MBG/silk composite scaffolds have also been prepared (see Figure 1.5) that show improved mechanical strength and excellent *in vitro* apatite-mineralization ability and *in vivo* osteogenesis (Wu, Zhang, Zhu, et al. 2010; Wu, Zhang, et al. 2011; Wu and Chang 2012).

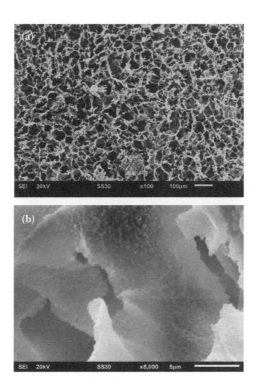

FIGURE 1.5
(a) MBG/silk composite scaffolds with (b) excellent apatite-mineralization ability in SBF.

1.3 Drug and Growth Factor Delivery of MBG

1.3.1 Controllable Delivery of Drug and Growth Factor in MBG

One of the significant advantages of MBG materials is that they possess higher specific surface area and pore volume than conventional bioactive glasses, which makes them useful for drug delivery. In the past several years, MBG with different material forms have been used for the study of drug delivery. Table 1.2 summarizes the drug and growth factor category for the different forms of MBG (Wu and Chang 2012). It can be seen that different categories of drug and growth factors (vascular endothelial growth factor [VEGF] and bone morphogenetic protein [BMP]) can be efficiently loaded and released in MBG particles, disks, spheres, fibers, scaffolds, and composites. The loading efficiency of drug and growth factors in MBG is significantly higher than that in conventional bioactive glasses (Xia and Chang 2006; Zhu and Kaskel 2009). The drug release kinetic in MBG is lower than that in conventional bioactive glasses. The loading efficiency and release kinetic can be controlled by adjusting the composition of MBG. Zhao, Yan, et al. (2008) found that the increase of CaO content in the MBG led to the enhancement of loading

TABLE 1.2

Drug/Growth Factor Category for MBG Materials

MBG Materials	Drug/Growth Factor Category	Reference
Particles	Gentamicin	Xia and Chang 2006, 2008
	Ibuprofen	Li, Wang, Hua, et al. 2008; Lin et al. 2009; Fan, Huang, et al. 2011
	Ipriflavone	Lopez-Noriega et al. 2010
Disks	Tetracycline	Zhao, Yan, et al. 2008
	Metoclopramide	Zhao, Loo, et al. 2008
	Phenanthrene	Lin et al. 2010
Spheres	Triclosan	Arcos et al. 2009
	Ibuprofen	Kang et al. 2011
	Ampicillin	Fan, Wu, et al. 2011
	Bovine serum albumin	Wu, Zhang, Ke, et al. 2010
Fibers	Gentamicin	Hong, Chen, Jing, Fan, Guo, et al. 2010; Hong, Chen, Jing, Fan, Gu, et al. 2010
Scaffolds	Gentamicin	Zhu and Kaskel 2009; Zhu, Li, et al. 2011; Zhu, Zhang, Wu, et al. 2011
	Dexamethasone	Wu, Zhang, Zhu, et al. 2010; Wu, Fan, Gelinsky, et al. 2011; Wu, Fan, Zhu, et al. 2011; Wu, Luo, et al. 2011; Wu, Miron, et al. 2011
	VEGF*	Wu et al., forthcoming
	BMP**	Dai et al. 2011
Composites	Gentamicin/naproxen	Xia et al. 2008; Li, Wang, et al. 2009; Zhu, Zhang, He, et al. 2011
	Dexamethasone	Wu, Ramaswamy, et al. 2009; Wu, Zhu, et al. 2010

Source: Wu, C., and Chang, J., 2012. Mesoporous bioactive glasses: Structure characteristics, drug/growth factor delivery and bone regeneration application. *Interface Focus* 2: 292–306.

* VEGF, vascular endothelial growth factor.

** BMP, bone morphogenetic protein.

efficiency and decrease of drug release rate and burst effect. The possible reason is that the drugs may be chelated with calcium on the pore wall, which makes it difficult to be released (Zhao, Yan, et al. 2008). It is found that 58S MBG has slower release kinetics than 77S MBG, indicating that CaO contents in MBG plays an important role in modulating the drug release (Xia and Chang 2006). Besides the composition, the mesoporous structure of MBG is of great importance to influence the drug delivery. Zhao, Loo, et al. (2008) used different surfactants (P123 and F127) to prepare MBG and found that P123-MBG had higher pore volume and surface area than F127-MBG. The higher pore volume and surface area of P123-MBG lead to significantly higher drug-loading efficiency (47.3%) compared to that of F127-MBG (16.6%). These studies suggest that MBG is an excellent drug and growth factor delivery system (Xia and Chang 2006; Wu, Miron, et al. 2011; Dai et al. 2011). The possible mechanism of slow drug release kinetics of MBG is due to the existence of a large number of Si-OH groups in MBG, which plays an important role in interacting with drugs and proteins (Xia and Chang 2006). Further study has found that the drug release from MBG is mainly controlled by a Fickian diffusion mechanism (Zhu and Kaskel 2009; Wu and Chang 2012).

Several methods were used to control drug and growth factor delivery in MBG sphere and scaffold systems. In MBG sphere system, we harnessed their apatite mineralization ability of MBG materials and coprecipitated bovine serum albumin (BSA) with apatite on the surface of MBG spheres. It was found that the BSA-loading efficiency of MBG was significantly enhanced by coprecipitating with apatite, and the loading efficiency and release kinetics of BSA could be controlled by controlling the density of apatite formed on MBG microspheres (Wu, Zhang, Ke, et al. 2010). Drug delivery could be controlled by using different bioactive materials with different degradation. Three bioactive material powders—MBG, nonmesoporous bioglass (BG), and hydroxyapatite (HAp)—were incorporated into alginate microspheres, respectively, and loaded with drugs. Results showed that the drug-loading capacity was enhanced with the incorporation of these materials into alginate microspheres. The MBG/alginate composite microspheres had the highest drug-loading capacity. Drug release from alginate microspheres correlated to the dissolution of MBG, BG, and HAp in PBS, and pH was a key factor in controlling the drug release; a high pH resulted in greater drug release, whereas a low pH delayed drug release (Wu, Zhu, et al. 2010). To further control drug delivery in MBG scaffold system, drugs were loaded into MBG scaffolds and then modified scaffolds by using silk protein, which formed a thin silk layer on the surface of pore walls. The silk layer can efficiently inhibit the burst release of drugs from MBG scaffolds (Wu, Zhang, Zhu, et al. 2010).

MBG scaffolds have been investigated for the delivery of VEGF. It was found that MBG scaffolds have significantly higher loading efficiency and more sustained release of VEGF than nonmesoporous bioactive glass scaffolds. Our study suggests that the mesopore structures in MBG scaffolds play an important role in improving the loading efficiency and decreasing the burst release of VEGF (Wu, Fan, et al., forthcoming).

1.3.2 Drug and Growth Factor Delivery in MBG for Antibacteria and Tissue-Stimulation Application

Although there are a great number of studies for the delivery of drugs by MBG, there are few studies for the functional effect of drug and growth factor delivery from MBG. Our group has recently investigated the antibacteria and cell-stimulating function of drug loaded MBG. Drug ampicillin was loaded into MBG nanospheres and scaffolds and then exposed to *E. coli* (DH5α) for different time periods. It was found that the sustained release of ampicillin from MBG revealed significant antibacteria effect (see Figure 1.6).

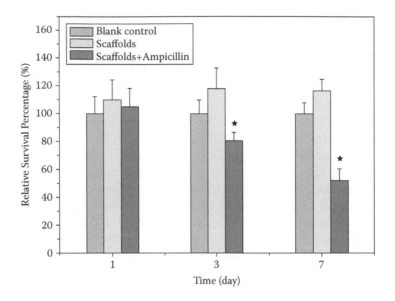

FIGURE 1.6
Sustained release of ampicillin from MBG scaffolds for their significant antibacteria effect. * Significant difference for the group of scaffolds loaded with ampicillin compared to blank control and scaffolds groups not loaded with ampicillin (P < 0.05).

Dexamethasone (DEX) was also loaded into MBG scaffolds, and it was found that the sustained release of DEX from MBG scaffolds significantly enhanced alkaline phosphatase (ALP) activity and gene expressions (ALP, BSP, and Col I) of osteoblasts. These results suggest that DEX-loaded MBG scaffolds show great potential as a release system to enhance osteogenesis and may be used for bone tissue engineering application (Wu, Miron, et al. 2011; Wu and Chang 2012).

The effect of VEGF delivery from MBG scaffolds on the viability of endothelial cells was further investigated and it was found that the mesopore structures in MBG scaffolds play an important role in maintaining the bioactivity of VEGF, further improving the viability of endothelial cells, indicating that MBG scaffolds are an excellent carrier of VEGF for stimulating angiogenesis.

Dai et al. (2011) incorporated recombinant human bone morphogenetic protein-2 (rhBMP-2) into MBG scaffolds and showed that the delivery of rhBMP-2 significantly promoted the *in vitro* osteogenic differentiation of bone marrow stromal cells and induced the ectopic bone formation in the thigh muscle pouches of mice. They further found that the delivery of rhBMP-2 resulted in more bone regeneration as compared to MBG scaffolds without rhBMP-2 (Dai et al. 2011). These studies suggest that MBG materials are a very useful carrier for drug and growth delivery with improved functions (Wu and Chang 2012).

1.4 Bone Tissue Engineering of MBG

1.4.1 Excellent Apatite-Mineralization Ability of MBG

One significant advantage of MBG is that it possesses superior apatite-mineralization ability in biological solution. The characteristics of MBG indicate that it may be used for bone regeneration applications. In the past several years, a great number of studies have focused on the *in vitro* bioactivity of MBG with different forms, including particles and scaffolds (Wu and Chang 2012). Yan et al. (2004) investigated the apatite mineralization of MBG particles in simulated body fluids and found that apatite formed only after soaking for 4 h. The apatite-mineralization ability of MBG was significantly higher than that of conventional nonmesoporous bioactive glasses. It is speculated that the high specific surface area and pore volume of MBG play an important role in the enhancement of the bioactive behavior of the MBG materials (Yan et al. 2004). They further optimized the composition structure–bioactivity correlation of MBG and found that the *in vitro* bioactivity of MBG is dependent on the Si to Ca ratio in the glass network. MBG (80Si15Ca5P) with relatively lower calcium content exhibits the best *in vitro* bioactivity in contrast to conventional melt-derived NBG where usually higher calcium percentage BG (e.g., 60Si35Ca) show better bioactivity (Yan et al. 2006). The mesopore size is also of importance to influence the apatite formation of MBG (Deng et al. 2009; Yu et al. 2010, 2011). Gunawidjaja et al. (2010) studied the mechanism of apatite mineralization of MBG by using nuclear magnetic resonance spectroscopy. It was found that the significant difference of the apatite formation mechanism between MBG and conventional NBG is that MBG does not require the typical "first 3 stages" (Gunawidjaja et al. 2010), but conventional NBG does (Hench and Polak 2002). In the first 3 stages, conventional NBG releases M^+ ions and form Si-OH groups and then Si-OH groups form networks by repolymerization. However, the surface of MBG is already inherently "prepared" to accelerate the first 3 stages of the conventional NBG (Gunawidjaja et al. 2010).

The apatite-mineralization ability of nanosized MBG particles, large-sized MBG beads, and a series of 3D MBG scaffolds with varied chemical compositions were investigated. All of them showed excellent apatite mineralization ability in biological solutions (Figure 1.7). The apatite-mineralization ability of MBG was significantly influenced by their chemical compositions. Typically, MBG with Ca-P-Si compositions showed optimal apatite mineralization ability. It was found that the incorporation of parts of Fe, Co, Sr, or Zr ions into MBG scaffolds could decrease their apatite-mineralization ability (Zhu and Kaskel 2009; Wu, Fan, Zhu, et al. 2011; Wu, Luo, et al. 2011; Wu, Miron, et al. 2011; Wu and Chang 2012); however, the incorporation of these ions into MBG scaffolds could enhance their *in vitro* cytocompatibility.

FIGURE 1.7
(a) Apatite mineralization on the surface of nanosized MBG particles, (b) large-sized MBG beads, and (c) porous MBG scaffolds.

1.4.2 *In Vitro* Cell Response to MBG

MBG has been regarded as a potential bioactive bone regeneration material due to its superior bioactive behavior. Further *in vitro* cell response to MBG materials has been studied. We compared the attachment and viability of human osteoblasts on four-composition MBG scaffolds and found that MBG scaffolds with the composition of Si80Ca15P5 had the best cell attachment (Zhu et al. 2008). MBG also supports the proliferation of human Saos-2 osteoblasts, murine L929 fibroblasts, and murine SR.D10 T lymphocytes (Alcaide et al. 2010). The incorporation of MBG particles into PLGA films enhanced the proliferation and ALP activity of human osteoblasts (Wu, Ramaswamy, et al. 2009). Our study further showed that silk modification of MBG scaffolds significantly improved the attachment, proliferation, differentiation, and osteogenic gene expression of bone marrow stromal cells (Wu, Zhang, Zhu, et al. 2010). In addition, the incorporation of Fe, Sr, B, and Zr ions into MBG scaffolds could not only support the attachment of BMSCs (Figure 1.8), but also enhance cell proliferation and osteogenic differentiation (Wu, Fan, Gelinsky, et al. 2011; Wu, Fan, Zhu, et al. 2011; Wu, Miron, et al. 2011; Zhu, Zhang, Wu, et al. 2011).

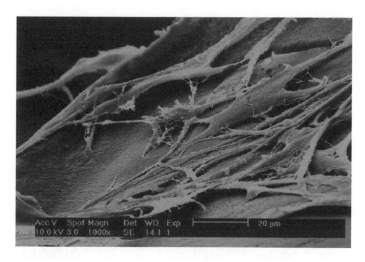

FIGURE 1.8
BMSCs morphology on MBG scaffolds after culturing for 7 days.

Our recent study revealed that the incorporation of Co^{2+} ions into MBG scaffolds significantly enhanced VEGF protein secretion, hypoxia-inducible factor (HIF)-1α expression, and VEGF gene expression of BMSCs. The incorporation of Co into MBG scaffolds is an efficient way to prepare hypoxia-mimicking tissue engineering scaffolds with significantly improved hypoxia function (Wu, Zhou, et al. 2012). The incorporation of Sr into MBG scaffolds has significantly stimulated the proliferation, ALP activity, and osteogenesis- and cementogenesis-related gene expression of periodontal ligament cells. The results suggested that bioactive ions released from MBG play an important role in enhancing cell response and their biological functions.

1.4.3 *In Vivo* Osteogenesis of MBG

To investigate the *in vivo* osteogenesis of MBG, MBG particles were implanted into the defects of rat femur. After 8 weeks of implantation, MBG particles induced a great amount of new bone ingrowths in the defects (see Figure 1.9). Furthermore, MBG particles were incorporated into silk scaffolds and investigated the *in vivo* osteogenesis. The study showed that MBG/silk scaffolds induced a higher rate of type I collagen synthesis and new bone formation after implanted in rat calvarial defects compared to conventional NBG/silk scaffolds. The results confirm that MBG has significant capacity to improve the *in vivo* bioactivity of silk scaffolds (Wu, Zhang, et al. 2011). The preliminary results indicate that MBG has excellent *in vivo* osteogenesis for potential bone repair application.

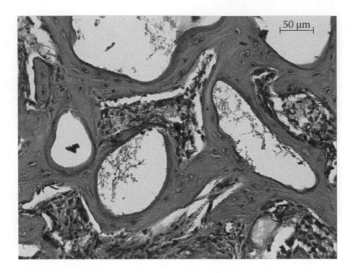

FIGURE 1.9
(See color insert.) New bone formation around the MBG particles after implanted into the defects of rat femur for 8 weeks (red area: new bone; white area: MBG particles).

1.5 Conclusion

In this chapter, the recent research advances of a new class of bioactive glasses named mesoporous bioactive glasses have been highlighted. We summarized the preparation methods, compositions, and their mesopore structures of different forms of MBG materials including particles, fibers, spheres, scaffolds, and composites. The main properties of MBG, such as drug delivery, apatite mineralization, *in vitro* cell response and *in vivo* osteogenesis, have been reviewed. From our studies, it was found that MBGs offer a suite of features that are important for efficient drug delivery and bone regeneration. The typical features include that (1) they possess well-ordered mesoporous channel structure and controllable nanopore size, which endows them high specific surface area and pore volume. The improved physical properties play an important role in the enhancement of their biological behavior; (2) they have excellent *in vitro* and *in vivo* bioactivity. Particularly, they have shown the stimulatory effect to enhance *in vivo* bone formation, compared to NBG; and (3) they are excellent carriers for the delivery of drugs and growth factors to further improve antibacteria ability and stimulate the growth and differentiation of tissue cells as well as bone tissue formation.

The important issues such as *in vivo* bone formation mechanism and MBG degradation process have to be investigated before it is ready for further clinical trials. If we can further understand the biological activity and corresponding mechanism of MBG *in vivo*, it is expected that MBG will be a promising biomaterial for clinical application in the future. For this aim, it is suggested that the potential directions for the study of MBG include (1) large animal models should be used to further investigate the *in vivo* osteogenesis and degradation of MBG; and (2) the effect of drug delivery on the *in vivo* osteogenesis is unclear, and further studies need to be conducted in the future.

Acknowledgments

The authors would like to thank the Shanghai Municipal Natural Science Foundation (12ZR1435300), the One Hundred Talent Project, SIC-CAS (Chengtie Wu), the Natural Science Foundation of China (Grant 81190132), and ARC Discovery (DP120103697) for the support of this research.

References

Alcaide, M., Portoles, P., Lopez-Noriega, A., et al., 2010. Interaction of an ordered mesoporous bioactive glass with osteoblasts, fibroblasts and lymphocytes, demonstrating its biocompatibility as a potential bone graft material. *Acta Biomater* 6: 892–9.

Arcos, D., Lopez-Noriega, A., Ruiz-Hernandez, E., et al., 2009. Ordered mesoporous microspheres for bone grafting and drug delivery. *Chem Mater* 21: 1000–1009.

Arcos, D., and Vallet-Regi, M., 2010. Sol-gel silica-based biomaterials and bone tissue regeneration. *Acta Biomater* 6: 2874–88.

Brinker, C. J., Lu, Y. F., Sellinger, A., et al., 1999. Evaporation-induced self-assembly: Nanostructures made easy. *Adv Mater* 11: 579–85.

Cerruti, M., and Sahai, N., 2006. Silicate biomaterials for orthopaedic and dental implants. *Rev Mineral Geochem* 64: 283–313.

Chen, Q. Z., Thompson, I. D., and Boccaccini, A. R., 2006. 45S5 Bioglass-derived glass-ceramic scaffolds for bone tissue engineering. *Biomaterials* 27: 2414–25.

Dai, C., Guo, H., Lu, J., et al., 2011. Osteogenic evaluation of calcium/magnesium-doped mesoporous silica scaffold with incorporation of rhBMP-2 by synchrotron radiation-based muCT. *Biomaterials* 32: 8506–17.

Deng, Y., Li, X. K., and Li, Q., 2009. Effect of pore size on the growth of hydroxyapatite from mesoporous CaO-SiO(2) substrate. *Ind Eng Chem Res* 48: 8829–36.

Fan, W., Wu, C., Han, P., et al., 2011. Porous Ca-Si-based nanospheres: A potential intra-canal disinfectant for infected canal treatment. *Mater Lett* 81: 16–19.

Fan, Y., Huang, S., Jiang, J., et al., 2011. Luminescent, mesoporous, and bioactive europium-doped calcium silicate (MCS: Eu3+) as a drug carrier. *J Colloid Interface Sci* 357: 280–5.

Feng, X., and Chang, J., 2011. Synthesis of a well-ordered mesoporous 58S bioactive glass by a simple method. *Int J Appl Ceram Technol* 8: 547–52.

Franco, J., Hunger, P., Launey, M. E., et al., 2010. Direct write assembly of calcium phosphate scaffolds using a water-based hydrogel. *Acta Biomater* 6: 218–28.

Garcia, A., Cicuendez, M., Izquierdo-Barba, I., et al., 2009. Essential role of calcium phosphate heterogeneities in 2D-hexagonal and 3D-cubic SiO2-CaO-P2O5 mesoporous bioactive glasses. *Chem Mater* 21: 5474–84.

Garcia, A., Izquierdo-Barba, I., Colilla, M., et al., 2011. Preparation of 3-D scaffolds in the SiO2-P2O5 system with tailored hierarchical meso-macroporosity. *Acta Biomater* 7: 1265–73.

Gough, J. E., Jones, J. R., and Hench, L. L., 2004. Nodule formation and mineralisation of human primary osteoblasts cultured on a porous bioactive glass scaffold. *Biomaterials* 25: 2039–46.

Gough, J. E., Notingher, I., and Hench, L. L., 2004. Osteoblast attachment and mineralized nodule formation on rough and smooth 45S5 bioactive glass monoliths. *J Biomed Mater Res A* 68: 640–50.

Gunawidjaja, P. N., Lo, A. Y. H., Izquierdo-Barba, I., et al., 2010. Biomimetic apatite mineralization mechanisms of mesoporous bioactive glasses as probed by multinuclear (31)P, (29)Si, (23)Na and (13)C solid-state NMR. *J Phys Chem C* 114: 19345–56.

Hamadouche, M., Meunier, A., Greenspan, D. C., et al., 2001. Long-term *in vivo* bioactivity and degradability of bulk sol-gel bioactive glasses. *J Biomed Mater Res* 54: 560–6.

Hench, L. L., 1991. Bioceramics: From concept to clinic. *J Am Ceram Soc* 74: 1487–1510.

Hench, L. L., 1998. Biomaterials: A forecast for the future. *Biomaterials* 19: 1419–23.

Hench, L. L., and Polak, J. M., 2002. Third-generation biomedical materials. *Science* 295: 1014–17.

Hench, L. L., and Thompson, I., 2010. Twenty-first century challenges for biomaterials. *J R Soc Interface* 7 Suppl 4: S379–391.

Hong, Y. L., Chen, X. S., Jing, X. B., Fan, H., Gu, Z., et al., 2010. Fabrication and drug delivery of ultrathin mesoporous bioactive glass hollow fibers. *Adv Funct Mater* 20: 1501–10.

Hong, Y., Chen, X., Jing, X., Fan, H., Guo, B., et al., 2010. Preparation, bioactivity, and drug release of hierarchical nanoporous bioactive glass ultrathin fibers. *Adv Mater* 22: 754–8.

Hoppe, A., Guldal, N. S., and Boccaccini, A. R., 2011. A review of the biological response to ionic dissolution products from bioactive glasses and glass-ceramics. *Biomaterials* 32: 2757–74.

Horcajada, P., Ramila, A., Boulahya, K., et al., 2004. Bioactivity in ordered mesoporous materials. *Solid State Sci* 6: 1295–1300.

Izquierdo-Barba, I., Ruiz-Gonzalez, L., Doadrio, J. C., et al., 2005. Tissue regeneration: A new property of mesoporous materials. *Solid State Sci* 7: 983–9.

Jones, J. R., Ehrenfried, L. M., and Hench, L. L., 2006. Optimising bioactive glass scaffolds for bone tissue engineering. *Biomaterials* 27: 964–73.

Jones, J. R., Tsigkou, O., Coates, E. E., et al., 2007. Extracellular matrix formation and mineralization on a phosphate-free porous bioactive glass scaffold using primary human osteoblast (HOB) cells. *Biomaterials* 28: 1653–63.

Kang, X., Huang, S., Yang, P., et al., 2011. Preparation of luminescent and mesoporous Eu3+/Tb3+ doped calcium silicate microspheres as drug carriers via a template route. *Dalton Trans* 40: 1873–9.

Lei, B., Chen, X. F., Wang, Y. J., and Zhao, N., 2009. Synthesis and *in vitro* bioactivity of novel mesoporous hollow bioactive glass microspheres. *Mater Lett* 63: 1719–21.

Lei, B., Chen, X. F., Wang, Y. J., Zhao, N., Du, C., et al., 2009. Acetic acid derived mesoporous bioactive glasses with an enhanced *in vitro* bioactivity. *J Non-Cryst Solid* 355: 2583–7.

Leonova, E., Izquierdo-Barba, I., Arcos, D., et al., 2008. Multinuclear solid-state NMR studies of ordered mesoporous bioactive glasses. *J Phys Chem C* 112: 5552–62.

Li, R., Clark, A. E., and Hench, L. L., 1991. An investigation of bioactive glass powders by sol-gel processing. *J Appl Biomater* 2: 231–9.

Li, X., Shi, J., Dong, X., et al., 2008. A mesoporous bioactive glass/polycaprolactone composite scaffold and its bioactivity behavior. *J Biomed Mater Res A* 84: 84–91.

Li, X., Wang, X. P., Chen, H. R., et al., 2007. Hierarchically porous bioactive glass scaffolds synthesized with a PUF and P123 cotemplated approach. *Chem Mater* 19: 4322–6.

Li, X., Wang, X. P., He, D. N., et al., 2008. Synthesis and characterization of mesoporous CaO-MO-SiO2-P2O5 (M = Mg, Zn, Cu) bioactive glasses/composites. *J Mater Chem* 18: 4103–9.

Li, X., Wang, X. P., Hua, Z. L., et al., 2008. One-pot synthesis of magnetic and mesoporous bioactive glass composites and their sustained drug release property. *Acta Mater* 56: 3260–5.

Li, X., Wang, X., Zhang, L., et al., 2009. MBG/PLGA composite microspheres with prolonged drug release. *J Biomed Mater Res B Appl Biomater* 89: 148–54.

Lin, H. M., Wang, W. K., Hsiung, P. A., et al., 2010. Light-sensitive intelligent drug delivery systems of coumarin-modified mesoporous bioactive glass. *Acta Biomater* 6: 3256–63.

Lin, J., Fan, Y., Yang, P. P., et al., 2009. Luminescent and mesoporous europium-doped bioactive glasses (MBG) as a drug carrier. *J Phys Chem C* 113: 7826–30.

Lopez-Noriega, A., Arcos, D., and Vallet-Regi, M., 2010. Functionalizing mesoporous bioglasses for long-term anti-osteoporotic drug delivery. *Chem Eur J* 16: 10879–86.

Lopez-Noriega, A., Arcos, D., Izquierdo-Barb, I., et al., 2006. Ordered mesoporous bioactive glasses for bone tissue regeneration. *Chem Mater* 18: 3137–44.

Manzanoab, M., and Vallet-Regi, M., 2010. New developments in ordered mesoporous materials for drug delivery. *J Mater Chem* 20: 5593–5604.

Miranda, P., Pajares, A., Saiz, E., et al., 2008. Mechanical properties of calcium phosphate scaffolds fabricated by robocasting. *J Biomed Mater Res A* 85: 218–27.

Miranda, P., Saiz, E., Gryn, K., et al., 2006. Sintering and robocasting of beta-tricalcium phosphate scaffolds for orthopaedic applications. *Acta Biomater* 2: 457–66.

Misra, S. K., Ansari, T., Mohn, D., et al., 2010. Effect of nanoparticulate bioactive glass particles on bioactivity and cytocompatibility of poly(3-hydroxybutyrate) composites. *J R Soc Interface* 7: 453–65.

Mourino, V., and Boccaccini, A. R., 2010. Bone tissue engineering therapeutics: Controlled drug delivery in three-dimensional scaffolds. *J R Soc Interface* 7: 209–27.

Salinas, A. J., Shruti, S., Malavasi, G., et al., 2011. Substitutions of cerium, gallium and zinc in ordered mesoporous bioactive glasses. *Acta Biomater* 7: 3452–8.

Valerio, P., Pereira, M. M., Goes, A. M., et al., 2004. The effect of ionic products from bioactive glass dissolution on osteoblast proliferation and collagen production. *Biomaterials* 25: 2941–8.

Vallet-Regi, M., 2006. Revisiting ceramics for medical applications. *Dalton Trans* 5211–20.

Vallet-Regi, M., Balas, F., and Arcos, D., 2007. Mesoporous materials for drug delivery. *Angew Chem Int Ed Engl* 46: 7548–58.

Vallet-Regi, M. A., Ruiz-Gonzalez, L., Izquierdo-Barba, I., et al., 2006. Revisiting silica based ordered mesoporous materials: Medical applications. *J Mater Chem* 16: 26–31.

Wei, J., Chen, F., Shin, J. W., et al., 2009. Preparation and characterization of bioactive mesoporous wollastonite—Polycaprolactone composite scaffold. *Biomaterials* 30: 1080–8.

Wu, C., and Chang, J., 2012. Mesoporous bioactive glasses: Structure characteristics, drug/growth factor delivery and bone regeneration application. *Interface Focus* 2: 292–306.

Wu, C., Chang, J., and Xiao, Y., 2011. Mesoporous bioactive glasses as drug delivery and bone tissue engineering platforms. *Therapeutic Delivery* 2: 1189–98.

Wu, C., Fan, W., Chang, J., et al., Forthcoming. Mesoporous bioactive glass scaffolds for efficient delivery of vascular endothelial growth factor. *J Biomater Appl.*

Wu, C., Fan, W., Gelinsky, M., et al., 2011. Bioactive SrO-SiO(2) glass with well-ordered mesopores: Characterization, physiochemistry and biological properties. *Acta Biomater* 7: 1797–806.

Wu, C., Fan, W., Zhu, Y., et al., 2011. Multifunctional magnetic mesoporous bioactive glass scaffolds with a hierarchical pore structure. *Acta Biomater* 7: 3563–72.

Wu, C., Luo, Y., Cuniberti, G., et al., 2011. Three-dimensional printing of hierarchical and tough mesoporous bioactive glass scaffolds with a controllable pore architecture, excellent mechanical strength and mineralization ability. *Acta Biomater* 7: 2644–50.

Wu, C., Miron, R., Sculeaan, A., et al., 2011. Proliferation, differentiation and gene expression of osteoblasts in boron-containing associated with dexamethasone deliver from mesoporous bioactive glass scaffolds. *Biomaterials* 32: 7068–78.

Wu, C., Ramaswamy, Y., Zhu, Y., et al., 2009. The effect of mesoporous bioactive glass on the physiochemical, biological and drug-release properties of poly(DL-lactide-co-glycolide) films. *Biomaterials* 30: 2199–208.

Wu, C., Zhang, Y., Ke, X., et al., 2010. Bioactive mesopore-glass microspheres with controllable protein-delivery properties by biomimetic surface modification. *J Biomed Mater Res A* 95: 476–85.

Wu, C., Zhang, Y., Zhou, Y., et al., 2011. A comparative study of mesoporous-glass/silk and non-mesoporous-glass/silk scaffolds: Physiochemistry and *in vivo* osteogenesis. *Acta Biomater* 7: 2229–36.

Wu, C., Zhang, Y., Zhu, Y., et al., 2010. Structure-property relationships of silk-modified mesoporous bioglass scaffolds. *Biomaterials* 31: 3429–38.

Wu, C., Zhou, Y., Fan, W., et al., 2012. Hypoxia-mimicking mesoporous bioactive glass scaffolds with controllable cobalt ion release for bone tissue engineering. *Biomaterials* 33: 2076–85.

Wu, C., Zhu, Y., Chang, J., et al., 2010. Bioactive inorganic-materials/alginate composite microspheres with controllable drug-delivery ability. *J Biomed Mater Res B Appl Biomater* 94: 32–43.

Wu, X., Wei, J., Lu, X., et al., 2010. Chemical characteristics and hemostatic performances of ordered mesoporous calcium-doped silica xerogels. *Biomed Mater* 5: 035006.

Wu, Z. Y., Hill, R. G., Yue, S., et al., 2011. Melt-derived bioactive glass scaffolds produced by a gel-cast foaming technique. *Acta Biomater* 7: 1807–16.

Xia, W., and Chang, J., 2006. Well-ordered mesoporous bioactive glasses (MBG): A promising bioactive drug delivery system. *J Control Release* 110: 522–30.

Xia, W., and Chang, J., 2008. Preparation, *in vitro* bioactivity and drug release property of well-ordered mesoporous 58S bioactive glass. *J Non-Cryst Solids* 15: 1338–41.

Xia, W., Chang, J., Lin, J., et al., 2008. The pH-controlled dual-drug release from mesoporous bioactive glass/polypeptide graft copolymer nanomicelle composites. *Eur J Pharm Biopharm* 69: 546–52.

Xynos, I. D., Edgar, A. J., Buttery, L. D., et al., 2000. Ionic products of bioactive glass dissolution increase proliferation of human osteoblasts and induce insulin-like growth factor II mRNA expression and protein synthesis. *Biochem Biophys Res Commun* 276: 461–5.

Yan, X., Huang, X., Yu, C., et al., 2006. The in-vitro bioactivity of mesoporous bioactive glasses. *Biomaterials* 27: 3396–403.

Yan, X., Yu, C., Zhou, X., et al., 2004. Highly ordered mesoporous bioactive glasses with superior *in vitro* bone-forming bioactivities. *Angew Chem Int Ed Engl* 43: 5980–4.

Yu, C. Z., Wei, G. F., Yan, X. X., et al., 2011. Synthesis and in-vitro bioactivity of mesoporous bioactive glasses with tunable macropores. *Micropo Mesopor Mat* 143: 157–65.

Yu, C. Z., Yan, X. X., Wei, G. F., et al., 2010. Synthesis and *in vitro* bioactivity of ordered mesostructured bioactive glasses with adjustable pore sizes. *Micropor Mesopor Mat* 132: 282–9.

Yun, H. S., Kim, S. E., and Hyeon, Y. T., 2007. Design and preparation of bioactive glasses with hierarchical pore networks. *Chem Comm* 2139–41.

Yun, H. S., Kim, S. E., and Hyun, Y. T., 2009. Preparation of bioactive glass ceramic beads with hierarchical pore structure using polymer self-assembly technique. *Mater Chem Phys* 115: 670–6.

Yun, H. S., Kim, S. H., Lee, S., et al., 2010. Synthesis of high surface area mesoporous bioactive glass nanospheres. *Mater Lett* 64: 1850–3.

Yun, H., Kim, S. E., Hyun, Y. T., et al., 2008. Hierarchically mesoporous-macroporous bioactive glasses scaffolds for bone tissue regeneration. *J Biomed Mater Res B Appl Biomater* 87: 374–80.

Zhao, D., Feng, J., Huo, Q., et al., 1998. Triblock copolymer syntheses of mesoporous silica with periodic 50 to 300 angstrom pores. *Science* 279: 548–52.

Zhao, L. Z., Yan, X. X., Zhou, X. F., et al., 2008. Mesoporous bioactive glasses for controlled drug release. *Micropor Mesopor Mat* 109: 210–5.

Zhao, S., Li, Y. B., and Li, D. X., 2010. Synthesis and *in vitro* bioactivity of CaO-SiO2-P2O5 mesoporous microspheres. *Microporous Mesoporous Mater* 135: 67–73.

Zhao, Y. F., Loo, S. C., Chen, Y. Z., et al., 2008. *In situ* SAXRD study of sol-gel induced well-ordered mesoporous bioglasses for drug delivery. *J Biomed Mater Res A* 85: 1032–42.

Zhong, J., and Greenspan, D. C., 2000. Processing and properties of sol-gel bioactive glasses. *J Biomed Mater Res* 53: 694–701.

Zhu, H., Wu, B., Feng, X., et al., 2011. Preparation and characterization of bioactive mesoporous calcium silicate-silk fibroin composite films. *J Biomed Mater Res B Appl Biomater* 98B: 330–41.

Zhu, M., Zhang, L. X., He, Q. J., et al., 2011. Mesoporous bioactive glass-coated poly(L-lactic acid) scaffolds: A sustained antibiotic drug release system for bone repairing. *J Mater Chem* 21: 1064–72.

Zhu, Y. F., and Kaskel, S., 2009. Comparison of the *in vitro* bioactivity and drug release property of mesoporous bioactive glasses (MBGs) and bioactive glasses (BGs) scaffolds. *Micropor Mesopor Mater* 118: 176–82.

Zhu, Y., Li, X., Yang, J., et al., 2011. Composition-structure-property relationship of the CaO-MxOy-SiO2-P2O5 (M = Zr, Mg, Sr) mesoporous bioactive glass (MBG) scaffolds. *J Mater Chem* 21: 9208–18.

Zhu, Y., Wu, C., Ramaswamy, Y., et al., 2008. Preparation, characterization and *in vitro* bioactivity of mesoporous bioactive glasses (MBGs) scaffolds for bone tissue engineering. *Micropor Mesopor Mat* 112: 494–503.

Zhu, Y., Zhang, Y., Wu, C., et al., 2011. The effect of zirconium incorporation on the physiochemical and biological properties of mesoporous bioactive glasses scaffolds. *Micropor Mesopor Mater* 143: 311–9.

2

Silicate-Based Bioactive Ceramics for Bone Regeneration Application

Chengtie Wu, Yin Xiao, and Jiang Chang

CONTENTS

2.1 Introduction

As populations in the developed world grow increasingly older, the demand for synthetic materials to replace and repair bone tissue lost from injury or disease significantly increases (Hench and Polak 2002; Hench and Thompson 2010). It is reported that around 60% of the bone graft substitutes available on the market involve ceramics. These mainly include calcium sulphate (plaster of Paris) (Georgiade et al. 1993), calcium phosphate (Ca-P) ceramics, and bioactive glasses (Hench and Wilson 1984). The common characteristics of these ceramics are that they possess osteogenic activity and can be used for bone regeneration application. Calcium sulphate (plaster of Paris) has had a long clinical history for use as a bone graft substitute in various skeletal sites including mandibular, craniofacial, and long bone defects (Georgiade et al. 1993). The drawback of this material is its high degradation rate, which may result in mismatching with new bone formation (Varlet and Dauchy

1983). Ca-P–based bioceramics, such as conventional hydroxyapatite (HAp) (El-Ghannam et al. 1997), β-tricalcium phosphate (β-TCP), and HAp/β-TCP composites, have been used as bone regeneration materials. However, a major drawback of these materials is the lack of sufficient mechanical strength, rendering them unsuitable as a load-bearing bone substitute (Hench 1998). Previous studies have shown that the sintered Ca-P–based bioceramics, especially HAp, lack full biodegradability after implantation (Ducheyne et al. 1993; Lu et al. 2002). Although β-TCP ceramics have been regarded as biodegradable materials, their degradation kinetic tends to be slow *in vitro* and *in vivo* (Ni et al. 2008; Xu et al. 2008; Wang et al. 2012). In addition, it is generally accepted that conventional sintered Ca-P ceramics lack osteoinductivity.

In the early 1970s, Hench and his colleagues developed a new class of biomaterials, SiO_2-CaO-Na_2O-P_2O_5 glasses 45S5. One of the significant characteristics of Ca-Si–based bioactive glasses is that they can induce the formation of HAp layer or hydroxyl carbonated apatite (HCA) similar to the mineral phase of bone on their surface in simulated body fluids (SBF) (Hench 1973; Hench and Paschall 1973). This class of Ca-Si–based glasses was able to osseointegrate with host bone (Hench 1991). Further studies have also shown that the Ca and Si containing ionic products released from the Bioglass 45S5 contribute to its bioactivity, as both Ca and Si are found to stimulate osteoblast proliferation and differentiation (Xynos et al. 2000; Gough, Jones, et al. 2004; Gough, Notingher, et al. 2004; Valerio et al. 2004; Hoppe et al. 2011). Xynos et al. (2000) using cDNA microarray analysis showed that the expression of a potent osteoblast mitogenic growth factor, insulin-like growth factor II (IGF-II), was increased to 32% upon exposure of the osteoblasts to the bioactive glass stimuli. Results indicated that Ca and Si ionic products from bioglass might increase IGF-II in osteoblasts by inducing the transcription of the growth factor and its carrier protein, and also by regulating the dissociation of this factor from the binding protein. The unbound IGF-II was likely to be responsible for the increase in cell proliferation observed in cell culture experiments. Similar bioactive induction of the transcription of extracellular matrix components and their secretion and self-organization into mineralized matrix might be responsible for the rapid formation and growth of bone nodules and differentiation of the mature osteocyte phenotype in the presence of bioglass (Hench et al. 2004). For this reason, Hench proposed the new concept of "third generation biomaterials" in which bioactive materials should stimulate cell and tissue growth. Recently, Hench suggested that bioactive glasses possess osteostimulation properties and they can be one of typical third generation biomaterials (Hench and Thompson 2010). However, the major disadvantages of bioglass are its high brittleness, low bending strength, fracture toughness, and workability, which limits its clinical application.

Silicon (Si) is one of the important trace elements in the human body. Si was found at a level of 100 ppm in bone and 200–550 ppm bound to extracellular matrix compounds (Schwarz 1973). It is reported that Si is located at active calcification sites in bone and is directly involved in the mineralization process of bone growth (Carlisle 1970). Inspired by the bioactive compositions of silicate-based bioglass and Si function in human body, a new family of bioactive silicate ceramics with a wide-range composition has been developed in the past 10 years. It was interesting to find that bioactive silicate ceramics with specific compositions could significantly stimulate *in vitro* osteogenic differentiation of several stem cells, and *in vivo* osteogenesis and angiogenesis. Bioactive silicate ceramics possess distinct osteostimulation properties. The study of bioactive silicate ceramics for bone tissue regeneration has been a hot area of research during the past several years. For this reason, it is of great importance to review the recent advances for bioactive silicate ceramics (not the silicate bioactive glasses). By summarizing the research progress in the field, the categories, mechanical properties, interaction with bone-forming cells, and *in vivo* osteogenesis and angiogenesis of silicate bioceramics have been mainly highlighted.

2.2 Preparation and Characterization of Silicate Bioceramics

2.2.1 Preparation of Silicate Bioceramics with Different Compositions

A series of silicate bioceramics have been prepared for bone regeneration and orthopedic coating application. As shown in Table 2.1, silicate bioceramics mainly include binary oxides ($CaO-SiO_2$, $MgO-SiO_2$, $SrO-SiO_2$, $ZnO-SiO_2$), ternary oxides ($MgO-CaO-SiO_2$, $ZnO-CaO-SiO_2$, $SrO-CaO-SiO_2$, $TiO_2-CaO-SiO_2$, $ZrO_2-CaO-SiO_2$, $P_2O_5-CaO-SiO_2$, $SrO-MgO-SiO_2$, $SrO-ZnO-SiO_2$, and $Na_2O-CaO-SiO_2$) and quaternary oxides ($SrO-ZnO-CaO-SiO_2$). Up to now, more than 20 silicate bioceramics with varied compositions have been prepared in the past 10 years. Compared to conventional phosphate-based bioceramics, silicate bioceramics have more broad chemical compositions, which may contribute to their adjustable physicochemical properties, such as mechanical strength, bioactivity, and degradability (Wu and Chang, forthcoming).

For the preparation of these silicate bioceramics, four main methods were applied for the synthesis of silicate powder materials, including the sol-gel method, chemical precipitation, hydrothermal method, and solid-reaction method. For binary-oxide silicate bioceramics, all the aforementioned methods can be used, compared to the synthesis of other composition-complex silicate powders, which often require more specific methods and more accurate control of the reaction conditions. By using the hydrothermal method, wollastonite ($CaSiO_3$) nanowires with high aspect ratios more than 100 and diameters of 50 to 100 nm were successfully prepared (Figure 2.1) (Li and Chang 2004). For preparation of pure ternary-oxide and quaternary-oxide

TABLE 2.1

Composition of Silicate Bioceramics

Name	Composition	Main Form	Reference
Binary Oxides			
Wollastonite	$CaSiO_3$	Powders, ceramics	De Aza et al. 1994, 1997, 1998, 1999, 2001; De Aza, Luklinska, Anseau, et al. 2000; De Aza, Luklinska, Martinez, et al. 2000; Ni et al. 2006; Xu et al. 2008
Dicalcium silicate	Ca_2SiO_4	Scaffolds	Gou et al. 2004; Gou and Chang 2004; Gou, Chang, and Zhai 2005; Gou, Chang, Zhai, et al. 2005; Huang and Chang 2007; Zhong et al. 2011
Tricalcium silicate	Ca_3SiO_5	Coatings	Zhao and Chang 2005; Zhao et al. 2005, 2007, 2008; Peng et al. 2011
Dimagnesium silicate	Mg_2SiO_4	Powders	Ni et al. 2007, 2008; Ni and Chang 2009; Kharaziha and Fathi 2010; Tavangarian and Emadi 2011a, 2011b, 2011c, 2011d
Magnesium silicate	$MgSiO_3$	Ceramics	Jin et al. 2011
Zinc silicate	Zn_2SiO_4	Ceramics	Zhang, Zhai, and Chang 2010
Strontium silicate	$SrSiO_3$	Powders	Zhang, Zhai, Lin et al., 2010
Ternary Oxides			
Akermanite	$Ca_2MgSi_2O_7$ $Ca_7MgSi_4O_{16}$	Powders	Wu and Chang 2004, 2006, 2007; Sun et al. 2006; Wu, Chang, Ni, et al. 2006; Wu, Chang, Zhai, et al. 2006; Liu et al. 2008; Huang et al. 2009; Gu et al. 2011; Xia et al. 2011; Zhai et al. 2012
Bredigite	$Ca_7MgSi_4O_{16}$	Ceramics	Wu, Chang, Wang et al. 2005; Wu and Chang 2007; Wu, Chang, et al. 2007; Huang and Chang 2008
Diopside	$CaMgSiO_4$	Spheres	Miake et al 1995; Nonami and Tsutsumi 1999; Wu and Chang 2007; Wu and Zreiqat 2010; Wu et al. 2010
Monticellite merwinite	$Ca_3MgSi_2O_8$ $Ca_3MgSi_2O_8$	Scaffolds	Chen et al. 2008; Ou et al. 2008; Hafezi-Ardakani et al. 2011
Hardystonite	$Ca_2ZnSi_2O_7$	Powders	Wu, Chang, and Zhai 2005; Ramaswamy, Wu, Zhou, et al. 2008; Lu et al. 2010
	$Zn_{(x)}CaSiO_{(3+x)}$	Ceramics	Wu, Ramaswamy, Chang, et al. 2008

TABLE 2.1 (*Continued*)

Composition of Silicate Bioceramics

Name	Composition	Main Form	Reference
	$(Sr,Ca)SiO_3$	Ceramics	Wu, Ramaswamy, et al. 2007; Zhang et al. 2011
Sphene	$CaTiSiO_5$	Ceramics, coatings	Wu, Ramaswamy, Soeparto, et al. 2008; Wu, Ramaswamy, Gale, et al. 2008; Ramaswamy et al. 2009; Wu et al. 2009
Baghdadite	$Ca_3ZrSi_2O_9$	Ceramics, spheres	Ramaswamy, Wu, Van Hummel, et al. 2008
Silicocarnotite	$Ca_5P_2SiO_{12}$	Powders, ceramics	Lu et al. 2012; Zhou et al. 2012
Nagelschmidtite	$Ca_7Si_2P_2O_{16}$		
Strontium-hardystonite	$Sr_2ZnSi_2O_7$	Ceramics	Zhang and Chang 2012
Calcium-sodium-silicate	$CaNa_2SiO_4$	Ceramics	Zhao et al. 2009
	$Ca_2Na_2Si_3O_9$		Du and Chang 2004
Quaternary Oxides			
	$(Sr,Ca)_2ZnSi_2O_7$	Scaffolds	Zreiqat et al. 2010

Source: Wu, C., and Chang, J., Forthcoming. Silicate bioceramics for bone tissue regeneration. *J Inorg Mater.*

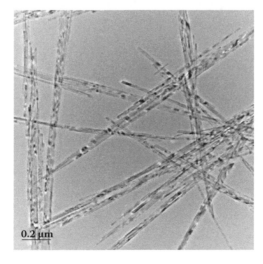

FIGURE 2.1
Silicate bioceramic nanowires synthesized by the hydrothermal method.

silicate powders, it is necessary to select proper chemicals, synthesis method, and control reaction conditions. To our knowledge, the sol-gel method is more suitable for the synthesis of the ternary-oxide and quaternary-oxide silicate powders than other methods (Wu and Chang, 2004, 2007, forthcoming; Wu, Ramaswamy, Soeparto, et al. 2008; Zreiqat et al. 2010). In addition, the calcining temperature for the sol-gel method is much lower than that for the solid-reaction method.

For preparation of dense silicate ceramic bulks, the conventional pressureless sintering technique was mostly selected. It is difficult to completely sinter silicate bioceramics with high density by using conventional sintering technique (see Figure 2.2) (Wu, Chang, Ni, et al. 2006). The main reason is that the calcining temperature to synthesize most of silicate powders is relatively high (>1000°C), which results in the crystal growth of silicate powders and in turn influences the density of ceramic bulks. We used a special sintering method, spark plasma sintering (SPS) (Long et al. 2006), to prepare dense silicate ceramics, and their density reached 99%.

Silicate bioceramics, such as wollastonite ($CaSiO_3$), diopside ($CaMgSi_2O_6$), akermanite ($Ca_2MgSi_2O_7$), bredigite ($Ca_7MgSi_4O_{16}$), and nagelschmidtite ($Ca_7Si_2P_2O_{16}$), have been prepared as three-dimensional (3D) porous scaffolds for bone tissue engineering applications (Lin et al. 2004; Ni et al. 2006; Wu, Chang, Zhai, et al. 2006; Wu, Chang, et al. 2007; Wu, Ramaswamy, Boughton, et al. 2008; Wu et al. 2010). There are three main techniques to prepare silicate bioceramic scaffolds. The first silicate bioceramic scaffold was prepared by the porogen method. Lin et al. (2004) prepared $CaSiO_3$ scaffolds by using polyethylene glycol (PEG) particulates. Although the prepared scaffolds

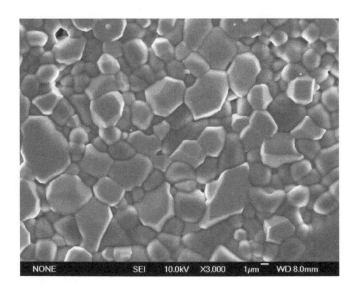

FIGURE 2.2
Akermanite ($Ca_2MgSi_2O_7$) bioceramics prepared by the pressureless sintering technique.

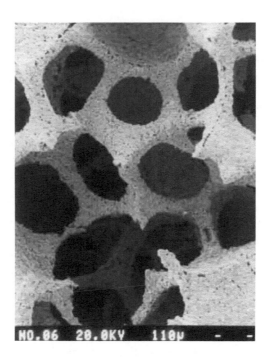

FIGURE 2.3
Silicate bioceramic scaffolds prepared by the polyurethane foam templating method.

have good mechanical strength, the main disadvantage of the $CaSiO_3$ scaffolds prepared by this method is that the pores are not uniform and interconnective (Lin et al. 2004), which is compromised for efficient cell infiltration and tissue ingrowths and nutrient transportation (Hutmacher 2000). We prepared $CaSiO_3$, $CaMgSi_2O_6$, $Ca_2MgSi_2O_7$, $Ca_7MgSi_4O_{16}$, and $Ca_7Si_2P_2O_{16}$ scaffolds with large pore size and high interconnectivity (see Figure 2.3) by using the polyurethane foam templating method; however, the main shortcoming of this method is that the prepared scaffolds are not mechanically strong (Ni et al. 2006). To better control the pore morphology, pore size, and porosity, a 3D plotting technique (also called direct writing or printing) has been developed to prepare porous silicate scaffolds. The significant advantage of this technique is that the architectures of the scaffolds can be concisely controlled by layer-by-layer plotting under mild conditions (Miranda et al. 2006, 2008; Franco et al. 2010). The $CaSiO_3$ scaffolds prepared by 3D plotting technique showed highly ordered large pore coordination (Wu et al. 2012; Wu and Chang, forthcoming).

2.2.2 Mechanical Strength of Silicate Bioceramics

The mechanical properties of bioactive materials are of great importance to influence their osteogenesis. It has been shown that the mechanical strength, especially the fracture toughness of some silicate bioceramic bulks, is obviously higher than that of conventional HAp ceramics, as shown in Table 2.2. Most silicate ceramic bulks possess a comparable bending strength and elastic modulus with human cortical bone (50–150 MPa) (Hench 1991). SPS-sintered silicate bioceramics, such as wollastonite and dicalcium silicate, have significantly improved mechanical strength compared to conventional pressureless sintered silicate ceramics (Long et al. 2006; Zhong et al. 2011). The bending strength of SPS-sintered silicate ceramics is higher than that of human cortical bone. The fracture toughness of SPS-sintered silicate ceramics is comparable to that of human cortical bone (Wu and Chang, forthcoming).

It is observed that $CaSiO_3$ scaffolds prepared by porogen methods have high compressive strength due to their low porosity and interconnectivity. The compressive strength of silicate ceramic scaffolds prepared by the polyurethane foam templating method is mainly in the range of 0.2 to 1.4 MPa, a relatively low mechanical strength. The compressive strength of 3D-plotted $CaSiO_3$ scaffolds is 3.6 MPa. The compressive strength of 3D-printed $CaSiO_3$ scaffolds is around 10 times that of polyurethane-templated $CaSiO_3$ scaffolds (Table 2.2). It was found that the mechanical profile of 3D-printed $CaSiO_3$ scaffolds increased almost linearly with the deformation of materials. Interestingly, after compressive testing, 3D-printed $CaSiO_3$ scaffolds partly maintain a scaffold configuration in the center position and only the border area collapses (Wu et al. 2012; Wu and Chang, forthcoming).

2.3 Physicochemical and Self-Setting Properties of Silicate Ceramics

After implantation of bioactive ceramics in the human body, a series of biochemical reactions occurred at the interface of the materials and host bone whereby a layer of bonelike apatite was formed on this interface, stimulating bone regeneration (Hench 1991). A first step in this process involved the release of Na^+/Ca^{2+} ions from bioactive ceramics and formation of Si-OH groups at the surface of materials. This was followed by Ca^{2+} and PO_4^{3-} ions being absorbed from body fluids forming an amorphous Ca-P deposition on the surface of ceramics. With the increase of implantation time, crystallized Ca-P (apatite) phase was formed. Finally, a matrix was produced stimulating the formation of new bone tissue. One possible way of investigating the *in vitro* bioactivity of bioceramics is by determining their ability to form an apatite layer in SBF. In this section, we introduce the interface chemistry reaction of ceramics with SBF solution and the apatite-formation mechanism on silicate ceramics in SBF. It is interesting to find that some silicate bioceramics possess distinct apatite-mineralization ability. Their apatite-mineralization

TABLE 2.2

Mechanical Strength of Silicate Bioceramics

Hydroxyapatite	Bending Strength (MPa)	Fracture Toughness (MPa·m$^{1/2}$)	Elastic Modulus (GPa)	Compressive Strength (MPa)	Reference
Dense Ceramic Bulks					
Hydroxyapatite	80–195	0.7–1.30	75–103		Hench 1998
Wollastonite	95				Lin et al. 2005
	294*	2.0*	46.5*		Long et al. 2006
Dicalcium silicate	26–97	1.1–1.8	10–40		Gou, Chang, and Zhai 2005
	293*	3.0*			Zhong et al. 2011
Tricalcium silicate	93.4	1.93	36.7		Zhao and Chang 2005
Calcium-silicate/ zirconia	395*	4.08*	81*		Long et al. 2008
Dimagnesium silicate	203	2.4			Ni et al. 2007
Magnesium silicate	32		8.5		Jin et al. 2011
Zinc silicate	91		37.5		Zhang, Zhai, and Chang 2010
Akermanite	176	1.83	42		Wu and Chang 2006
Bredigite	156	1.57	43		Wu, Chang, Wang, et al. 2005
Diopside	300	3.50			Nomami and Tsutsumi 1999
Merwinite	151	1.72	31		Ou et al. 2008
Monticellite	159	1.63	51		Chen et al. 2008
Hardystonite	136	1.37	37		Wu, Chang, and Zhai 2005
Silicocarnotite	65		80		Lu et al. 2012
Porous Scaffolds					
Wollastonite				60[a]	Lin et al. 2004
				0.4[b]	Ni et al. 2006
				3.6[c]	
Akermanite				0.53–1.13[b]	Wu, Chang, Zhai, et al. 2006
Diopside				0.2–1.36[b]	Wu et al. 2010
Bredigite				0.233[b]	Wu, Chang, et al. 2007

Source: Wu, C., and Chang, J., Forthcoming. Silicate bioceramics for bone tissue regeneration. *J Inorg Mater*.

* SPS sintering technique.

[a] Porogen method.

[b] Polyurethane foam templating method.

[c] 3D plotting technique.

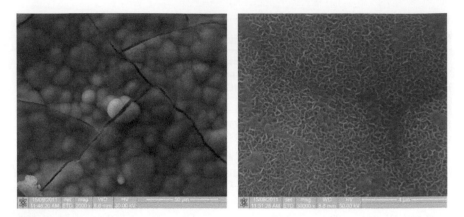

FIGURE 2.4
Apatite mineralization on silicate bioceramics with lath-like morphology in SBF.

ability was directly dependent on their chemical compositions and dissolution. Wollastonite, dicalcium silicate, tricalcium silicate, bredigite, and nagelschmidtite ceramics possess the best apatite-mineralization ability and quickest dissolution in SBF. Akermanite, merwinite, silicocarnotite, and strontium silicate have good apatite mineralization and moderate dissolution. Hardystonite bioceramics have no obvious apatite mineralization and their dissolution is quite low. The morphology of the formed apatite on typical silicate ceramics is shown in Figure 2.4. Generally, silicate bioceramics with high Ca contents possess improved apatite mineralization in SBF. The incorporation of other metal ions, such as Mg, Zn, and Zr, into binary oxide silicate ceramics will decrease their apatite mineralization. Furthermore, it is found that the dissolution may play an important role to influence the apatite mineralization of silicate ceramics. Silicate ceramics with quick dissolution possess improved apatite mineralization ability, compared to those with slow dissolution. Our studies have shown that the apatite formation of silicate bioceramics may involve several steps. First, the ion exchange of Ca^{2+} and other metal ions in the silicate ceramics with H^+ in SBF resulted in the formation of a hydrated silica layer on the surfaces of the ceramics and provided favorable sites for phosphate nucleation (Wu, Chang, Wang, et al. 2005). Then, Ca^{2+} and PO_4^{3-} ions in SBF will remineralize and form amorphous calcium phosphate and the subsequent formation of crystal apatite by incorporating OH^- ions from SBF. The mechanism of apatite formation on the surfaces of silicate bioceramics is similar to that of 45S5 bioglass (Wu, Chang, Wang, et al. 2005; Wu and Chang, forthcoming).

Besides the apatite-mineralization properties of silicate bioceramics, it was found that some silicate ceramics, such as dicalcium silicate and tricalcium silicate, possessed a self-setting property and injectability. The self-setting property of silicate paste is due to the progressive hydration of the SiO_4^{4-} ions in silicate ceramics. When silicate ceramics react with water, a nanoporous, amorphous CSH gel is deposited on the original silicate materials, while $Ca(OH)_2$ crystals nucleate and grow in the available capillary pore space and the previously deposited CSH gel is well accepted. As time proceeds, the CSH gels polymerize and harden. The self-setting progress of silicate materials is mainly attributed to the formation of a solid network, which is also associated with the densification and increase of the mechanical strength (Gou, Chang, Zhai, et al. 2005; Zhao et al. 2005). Our group has developed a series of composite bone cements based on silicate and phosphate/$CaSO_4$ materials, which exhibit improved mechanical strength and setting properties for potential bone regeneration application (Huan and Chang 2007a,b, 2008, 2009; Zhao et al. 2008; Wu and Chang, forthcoming).

2.4 Biological Response of Bone-Forming Cells to Silicate Ceramics

2.4.1 Cell Attachments

The attachment, adhesion, and spreading belong to the first phase of cell–materials interactions and influence the cell's capacity to proliferate and differentiate on contact with the implant (Anselme 2000). In the past several years, we have conducted a series of cell experiments to evaluate cell attachment on more than 20 silicate bioceramics. It is found that most of silicate bioceramics support the attachment of osteoblast, bone marrow stromal cells (BMSCs) and periodontal ligament cells (PDLCs), such as akermanite (see Figure 2.5) (Wu, Chang, Ni, et al. 2006; Wu and Chang 2007; Wu, Ramaswamy, Chang, et al. 2008; Wu, Ramaswamy, Soeparto, et al. 2008). For bone tissue engineering application, osteoblasts have been found well attached on the pore walls of porous silicate bioceramic scaffolds (see Figure 2.6) (Wu, Chang, Zhai, et al. 2006; Wu et al. 2010). Generally, it is found that silicate bioceramics and scaffolds, such as akermanite, diopside, and hardystonite, with moderate or slow degradation benefit cell attachments, compared to those with quick degradation, such as wollastonite, bredigite, and tricalcium silicate. Therefore, it is speculated that the chemical composition of silicate bioceramics is the main reason to influence cell attachment. Their chemical compositions directly decide their surface degradation and further influence cell attachment (Wu and Chang, forthcoming).

2.4.2 *In Vitro* Osteostimulation of Silicate Bioceramics

There is a common characteristic for silicate bioceramics, which the Si-containing ionic products from silicate bioceramics significantly stimulate the proliferation and osteogenic differentiation of several kinds of stem cells,

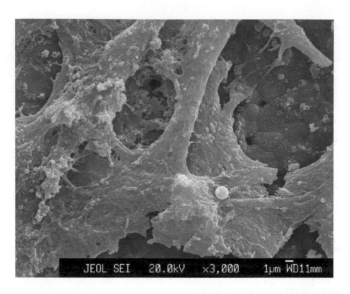

FIGURE 2.5
Osteoblast attachment on akermanite silicate bioceramics.

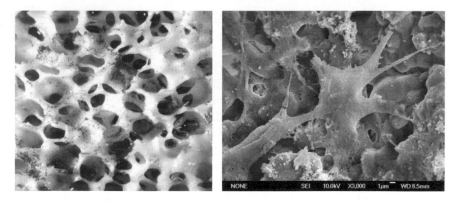

FIGURE 2.6
Osteoblast growing on porous akermanite silicate ceramic scaffolds for a bone tissue engineering application. Left: Optical micrograph. Right: SEM micrograph.

including dental pulp cells, bone marrow stromal cells, adipose-derived stem cells, and periodontal ligament cells, suggesting that silicate ceramics possess excellent osteostimulation properties *in vitro*. Akermanite showed the most distinct osteostimulation properties, which they can stimulate the proliferation for five kinds of cells and osteogenic differentiation for four stem cells (Sun et al. 2006; Wu, Chang, Ni, et al. 2006; Wu and Chang 2007; Gu et al. 2011; Xia et al. 2011). We found that the Ca, Si, and Mg ions from akermanite dissolution at certain ranges of concentration significantly stimulated osteoblast and L929 cell proliferation (Wu, Chang, Ni, et al. 2006). Further study showed that the ionic products from akermanite promoted the proliferation of hBMSC significantly more than did β-TCP extract. Alkaline phosphatase (ALP) activity and the expression of osteogenic marker genes of BMSCs, such as ALP, osteopontin (OPN), osteocalcin (OCN), and bone sialoprotein (BSP), were significantly enhanced by akermanite extract compared to β-TCP extract, indicating the osteogenic stimulation for akermanite (see Figure 2.7) (Huang et al. 2009). Besides the stimulatory effect on BMSCs, Ca, Mg, and Si ions extracted from akermanite in the concentrations of 2.36, 1.11, and 1.03 mM could facilitate the osteogenic differentiation of adipose-derived stem

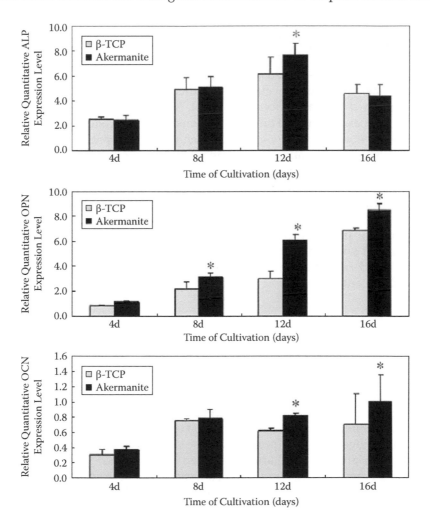

FIGURE 2.7
Bone-related gene expression of hBMSC was enhanced by akermanite extracts compared to that by β-TCP extracts. (ALP: alkaline phosphatase; OCN: osteocalcin; OPN: osteopontin.)

cells via an ERK pathway (Gu et al. 2011). Further study suggested that more pronounced proliferation and higher osteogenic gene expression for hPDLCs cultured with akermanite extract were detected as compared to cells cultured with β-TCP extract (Xia et al. 2011; Wu and Chang, forthcoming).

Akermanite could not only stimulate cell proliferation and osteogenic differentiation of several stem cells cultured with their ionic extracts, but also promote osteogenic differentiation of stem cells directly cultured with their ceramic disks (Sun et al. 2006; Wu, Chang, Ni, et al. 2006; Wu and Chang 2007; Gu et al. 2011; Xia et al. 2011). Therefore, it is suggested that silicate bioceramics possess significant osteostimulation properties (Wu and Chang, forthcoming).

To further investigate the potential mechanism of the osteostimulation properties of silicate bioceramics, the effect of Si ions on the proliferation, differentiation, bone-related gene expression, and cell signaling pathways of bone marrow stromal cells was further studied by comparing BMSC response to different concentrations of NaCl and Na_2SiO_3, taking into account and excluding the effect of Na ions. Our study showed that Si ions at a certain concentration significantly enhanced the proliferation, mineralization nodule formation, and bone-related gene expression (*OCN, OPN,* and *ALP*) of BMSCs. Furthermore, Si ions at 0.625 mM could counteract the effect of Wnt inhibitor on the osteogenic gene expression, (*OPN, OCN,* and *ALP*), Wnt and Shh signaling pathway-related genes in BMSCs. The results suggest that Si ions by themselves play an important role in regulating the proliferation and osetogenic differentiation of BMSCs with the involvement of Wnt and Shh signaling pathways (unpublished data). Therefore, Si ions released from silicate bioceramics can acquire enhanced bioactivity at desired concentration.

2.5 *In Vivo* Bone-Forming Ability of Silicate Bioceramics

Since some silicate bioceramics, such as wollastonite ($CaSiO_3$), akermanite ($Ca_2MgSi_2O_7$), and baghdadite ($Ca_3ZrSi_2O_9$), have shown excellent *in vitro* bioactivity, we conducted further *in vivo* studies to investigate their osteogenesis and angiogenesis. The porous wollastonite scaffolds were implanted in rabbit calvarial defects and after implanted for 4, 8, and 16 weeks, the micro-CT and histomorphometric analysis showed that the resorption of wollastonite scaffolds was much higher than that of β-TCP scaffolds. More bone formation was observed with wollastonite scaffolds as compared with β-TCP scaffolds (Xu et al. 2008). Our recent study demonstrated that the incorporation parts of wollastonite into β-TCP scaffolds could significantly promote new bone formation after implanted in critical sized femur defects of rabbits (Wang et al. 2012).

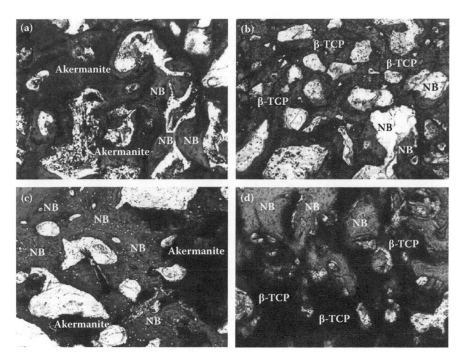

FIGURE 2.8
(See color insert.) High magnification images of new bone formation and material degradation of (a, c) akermanite and (b, c) β-TCP implants after (a, b) 8 and (c, d) 16 weeks (Van Gieson's
picrofuchsin staining of transverse section; NB: new bone). Red color indicates newly formed
bone. Original magnification: 100×.

Akermanite scaffolds were implanted into rabbit femur defect models and
the results indicated that both in early- and late-stage implantations, akermanite promoted more osteogenesis and biodegradation than did β-TCP
(Wu and Chang, forthcoming); and in late-stage implantations, the rate of
new bone formation was faster in akermanite than in β-TCP as shown in
Figure 2.8 (Huang et al. 2009). The akermanite ion extract predominantly
promoted the proliferation of human aortic endothelial cells and upregu-
lated the expression of genes encoding the receptors of proangiogenic
cytokines and the expression level of genes encoding the proangiogenic
downstream cytokines, such as nitric oxide synthase and nitric oxide synthesis. Akermanite implanted in the rabbit femoral condyle model promoted
neovascularization after 8 and 16 weeks of implantation, which confirmed
its stimulation effect on angiogenesis *in vivo* (Zhai et al. 2012).

Recently, 1-mm baghdadite ceramic spheres were implanted into the supra-
condylar site of the femur defects in Wistar rats and the degree of *in vivo*
osteogenesis was evaluated by hematoxylin and eosin, Safranin O staining,
tartrateresistant acid phosphatase (TRAP) staining, and immunohistochem-
istry (type I collagen: Col I; osteopontin: OPN) analyses. The results have

shown that baghdadite spheres induced a higher rate of new bone formation in the defects than did β-TCP spheres. Immunohistochemical analysis showed greater expression of Col I and OPN in the baghdadite group compared to the β-TCP groups (Luo et al. 2012). The results indicate that silicate bioceramics possess excellent *in vivo* osteostimulation. It is suggested that there are two main factors to contribute to the excellent *in vivo* osteogenesis of silicate bioceramics. One is that the released Si ions from silicate bioceramics stimulate osteogenic differentiation for bone-forming cells and further promote *in vivo* osteogenesis and angiogenesis; the other is that their excellent apatite-mineralization ability may contribute to their *in vivo* bioactivity and in turn stimulate *in vivo* osteogeneis.

2.6 Conclusion

This chapter reviewed the research advancement of silicate system bioceramics, including preparation methods, mechanical strength, apatite mineralization, dissolution, and *in vitro* and *in vivo* osteostimulation properties. Their osteostimulation properties and corresponding mechanism were highlighted. By comparing silicate bioceramics with conventional calcium phosphate bioceramics, it is found that some silicate bioceramics have superior physicochemical and osteostimulation properties. Silicate bioceramics possess two significant features: (1) silicate bioceramics could promote osteogenic differentiation of several kinds of stem cells, including bone marrow stromal cells, adipose stem cells, dental pulp, and periodontal cells; and (2) silicate bioceramics could induce *in vitro* and *in vivo* angiogenesis. It is known that how to improve the osteogenesis and angiogenesis of biomaterials remain a significant challenge; silicate bioceramics may provide a new way to solve these problems. It is widely accepted that the development of the third generation of biomaterials is a common goal for tissue regeneration application, which biomaterials in the human physiological environment should stimulate the functional response to tissue cells at the molecular level response and further promote new bone tissue formation. Silicate bioceramics have been confirmed to activate cell response at the gene level and promote new bone formation. Therefore, it is reasonable to regard silicate bioceramics as a typical representative of the third generation of bioactive materials. Moreover, since silicate bioceramics have a relatively wide range of chemical composition, their physical, chemical, and biological properties could be well optimized to satisfy the requirements of tissue regeneration.

We believe that some silicate bioceramics are promising for clinical application in 5 years. To achieve this goal, more studies are needed: (1) the molecular mechanism of osteostimulation by silicate bioceramics should be further clarified by culturing stem cells; and (2) the *in vivo* osteogenesis and angiogenesis of silicate bioceramics as well as the corresponding mechanism should be investigated in large animal models.

Acknowledgments

The authors thank the Shanghai Municipal Natural Science Foundation (12ZR1435300), One Hundred Talent Project, SIC-CAS, the Natural Science Foundation of China (Grant 81190132), and ARC Discovery (DP120103697) for research support.

References

Anselme, K., 2000. Osteoblast adhesion on biomaterials. *Biomaterials* 21: 667–81.
Carlisle, E. M., 1970. Silicon: A possible factor in bone calcification. *Science* 167: 279–80.
Chen, X., Ou, J., Kang, Y., et al., 2008. Synthesis and characteristics of monticellite bioactive ceramic. *J Mater Sci Mater Med* 19: 1257–63.
De Aza, P. N., Guitian, F., and De Aza, S., 1994. Bioactivity of wollastonite ceramics *in-vitro* evaluation. *Scrip Metal Et Mater* 31: 1001–5.
De Aza, P. N., Guitian, F., and De Aza, S., 1997. Bioeutectic: A new ceramic material for human bone replacement. *Biomaterials* 18: 1285–91.
De Aza, P. N., Guitian, F., De Aza, S., et al., 1998. Analytical control of wollastonite for biomedical applications by use of atomic absorption spectrometry and inductively coupled plasma atomic emission spectrometry. *Analyst* 123: 681–5.
De Aza, P. N., Luklinska, Z. B., Anseau, M. R., et al., 1999. Bioactivity of pseudowollastonite in human saliva. *J Dent* 27: 107–13.
De Aza, P. N., Luklinska, Z. B., Anseau, M. R., et al., 2000. Reactivity of a wollastonite-tricalcium phosphate bioeutectic ceramic in human parotid saliva. *Biomaterials* 21: 1735–41.
De Aza, P. N., Luklinska, Z. B., Anseau, M. R., et al., 2001. Transmission electron microscopy of the interface between bone and pseudowollastonite implant. *J Microsc* 201: 33–43.
De Aza, P. N., Luklinska, Z. B., Martinez, A., et al., 2000. Morphological and structural study of pseudowollastonite implants in bone. *J Microsc* 197: 60–7.
Du, R., and Chang, J., 2004. Preparation and characterization of bioactive sol-gel-derived Na2Ca2Si3(O)9. *J Mater Sci Mater Med* 15: 1285–9.
Ducheyne, P., Radin, S., and King, L., 1993. The effect of calcium phosphate ceramic composition and structure on *in vitro* behavior. I. Dissolution. *J Biomed Mater Res* 27: 25–34.
El-Ghannam, A., Ducheyne, P., and Shapiro, I. M., 1997. Formation of surface reaction products on bioactive glass and their effects on the expression of the osteoblastic phenotype and the deposition of mineralized extracellular matrix. *Biomaterials* 18: 295–303.

Franco, J., Hunger, P., Launey, M. E., et al., 2010. Direct write assembly of calcium phosphate scaffolds using a water-based hydrogel. *Acta Biomater* 6: 218–28.

Georgiade, N. G., Hanker, J., Levin, S., et al., 1993. The use of particulate hydroxyapatite and plaster of Paris in aesthetic and reconstructive surgery. *Aesthetic Plast Surg* 17: 85–92.

Gou, Z. G., and Chang, J., 2004. Synthesis and *in vitro* bioactivity of dicalcium silicate powders. *J Euro Ceram Soc* 24: 93–99.

Gou, Z. R., Chang, J., and Zhai, W. Y., 2005. Preparation and characterization of novel bioactive dicalcium silicate ceramics. *J Euro Ceram Soc* 25: 1507–14.

Gou, Z. G., Chang, J., Zhai, W. Y., et al., 2005. Study on the self-setting property and the *in vitro* bioactivity of beta-Ca2SiO4. *J Biomed Mater Res B-App Biomater* 73B: 244–51.

Gou, Z. R., Chang, J., Gao, J. H., et al., 2004. *In vitro* bioactivity and dissolution of Ca-2(SiO3)(OH)(2) and beta-Ca2SiO4 fibers. *J Euro Ceram Soc* 24: 3491–7.

Gough, J. E., Jones, J. R., and Hench, L. L., 2004. Nodule formation and mineralisation of human primary osteoblasts cultured on a porous bioactive glass scaffold. *Biomaterials* 25: 2039–46.

Gough, J. E., Notingher, I., and Hench, L. L., 2004. Osteoblast attachment and mineralized nodule formation on rough and smooth 45S5 bioactive glass monoliths. *J Biomed Mater Res A* 68: 640–50.

Gu, H., Guo, F., Zhou, X., et al., 2011. The stimulation of osteogenic differentiation of human adipose-derived stem cells by ionic products from akermanite dissolution via activation of the ERK pathway. *Biomaterials* 32: 7023–33.

Hafezi-Ardakani, M., Moztarzadeh, F., Rabiee, M., et al., 2011. Synthesis and characterization of nanocrystalline merwinite (Ca(3)Mg(SiO(4))(2)) via sol-gel method. *Ceram Inter* 37: 175–80.

Hench, L. L., 1973. Ceramics, glasses, and composites in medicine. *Med Instrum* 7: 136–44.

Hench, L. L., 1991. Bioceramics: From concept to clinic. *J Am Ceram Soc* 74: 1487–1510.

Hench, L. L., 1998. Biomaterials: A forecast for the future. *Biomaterials* 19: 1419–23.

Hench, L. L., and Paschall, H. A., 1973. Direct chemical bond of bioactive glass-ceramic materials to bone and muscle. *J Biomed Mater Res* 7: 25–42.

Hench, L. L., and Polak, J. M., 2002. Third-generation biomedical materials. *Science* 295: 1014–7.

Hench, L. L., and Thompson, I., 2010. Twenty-first century challenges for biomaterials. *J R Soc Interface* 7 Suppl 4: S379–91.

Hench, L. L., and Wilson, J., 1984. Surface-active biomaterials. *Science* 226: 630–6.

Hench, L. L., Xynos, I. D., and Polak, J. M., 2004. Bioactive glasses for *in situ* tissue regeneration. *J Biomater Sci Polym Ed* 15: 543–62.

Hoppe, A., Guldal, N. S., and Boccaccini, A. R., 2011. A review of the biological response to ionic dissolution products from bioactive glasses and glass-ceramics. *Biomaterials* 32: 2757–74.

Huan, Z., and Chang, J., 2007a. Novel tricalcium silicate/monocalcium phosphate monohydrate composite bone cement. *J Biomed Mater Res B Appl Biomater* 82: 352–9.

Huan, Z., and Chang, J., 2007b. Self-setting properties and *in vitro* bioactivity of calcium sulfate hemihydrate-tricalcium silicate composite bone cements. *Acta Biomater* 3: 952–60.

Huan, Z., and Chang, J., 2008. Study on physicochemical properties and *in vitro* bio-activity of tricalcium silicate-calcium carbonate composite bone cement. *J Mater Sci Mater Med* 19: 2913–8.

Huan, Z., and Chang, J., 2009. Novel bioactive composite bone cements based on the beta-tricalcium phosphate-monocalcium phosphate monohydrate composite cement system. *Acta Biomater* 5: 1253–64.

Huang, X. H., and Chang, J., 2007. Low-temperature synthesis of nanocrystalline beta-dicalcium silicate with high specific surface area. *J Nanoparticle Res* 9: 1195–1200.

Huang, X. H., and Chang, J., 2008. Preparation of nanocrystalline bredigite powders with apatite-forming ability by a simple combustion method. *Mater Res Bull* 43: 1615–20.

Huang, Y., Jin, X., Zhang, X., et al., 2009. *In vitro* and *in vivo* evaluation of akermanite bioceramics for bone regeneration. *Biomaterials* 30: 5041–8.

Hutmacher, D. W., 2000. Scaffolds in tissue engineering bone and cartilage. *Biomaterials* 21: 2529–43.

Jin, X. G., Chang, J. A., Zhai, W. Y., et al., 2011. Preparation and characterization of clinoenstatite bioceramics. *J Am Ceram Soc* 94: 173–7.

Kharaziha, M., and Fathi, M. H., 2010. Improvement of mechanical properties and biocompatibility of forsterite bioceramic addressed to bone tissue engineering materials. *J Mech Behav Biomed Mater* 3: 530–7.

Li, X. K., and Chang, J., 2004. Synthesis of wollastonite single crystal nanowires by a novel hydrothermal route. *Chem Lett* 33: 1458–9.

Lin, K. L., Chang, J., Zeng, Y., et al., 2004. Preparation of macroporous calcium silicate ceramics. *Mater Lett* 58: 2109–13.

Lin, K. L., Zhai, W. Y., Ni, S. Y., et al., 2005. Study of the mechanical property and *in vitro* biocompatibility of CaSiO3 ceramics. *Ceram Inter* 31: 323–6.

Liu, Q., Cen, L., Yin, S., et al., 2008. A comparative study of proliferation and osteo-genic differentiation of adipose-derived stem cells on akermanite and beta-TCP ceramics. *Biomaterials* 29: 4792–9.

Long, L. H., Chen, L. D., Bai, S. Q., et al., 2006. Preparation of dense beta-CaSiO3 ceramic with high mechanical strength and HAp formation ability in simulated body fluid. *J Euro Ceram Soc* 26: 1701–6.

Long, L. H., Zhang, F. M., Chen, L., et al., 2008. Preparation and properties of beta-CaSiO3/ZrO2 (3Y) nanocomposites. *J Euro Ceram Soc* 28: 2883–7.

Lu, H., Kawazoe, N., Tateishi, T., et al., 2010. *In vitro* proliferation and osteogenic dif-ferentiation of human bone marrow-derived mesenchymal stem cells cultured with hardystonite (Ca2ZnSi 2O7) and {beta}-TCP ceramics. *J Biomater Appl* 25: 39–56.

Lu, J. X., Descamps, M., Dejou, J., et al., 2002. The biodegradation mechanism of cal-cium phosphate biomaterials in bone. *J Biomed Mater Res* 63: 408–12.

Lu, W., Duan, W., Guo, Y., et al., 2012. Mechanical properties and *in vitro* bioactivity of Ca5(PO4)2SiO4 bioceramic. *J Biomater Appl* 26: 637–50.

Luo, T., Wu, C., and Zhang, Y., 2012. The *in vivo* osteogenesis of Mg- or Zr-modified silicate-based bioceramic spheres. *J Biomed Mater Res A* 2269–77.

Miake, Y., Yanagisawa, T., Yajima, Y., et al., 1995. High-resolution and analytical elec-tron microscopic studies of new crystals induced by a bioactive ceramic (diop-side). *J Dent Res* 74: 1756–63.

Miranda, P., Pajares, A., Saiz, E., et al., 2008. Mechanical properties of calcium phos-phate scaffolds fabricated by robocasting. *J Biomed Mater Res A* 85: 218–27.

Miranda, P., Saiz, E., Gryn, K., et al., 2006. Sintering and robocasting of beta-trical-cium phosphate scaffolds for orthopaedic applications. *Acta Biomater* 2: 457–66.

Ni, S. Y., Chou, L., and Chang, J., 2007. Preparation and characterization of forsterite (Mg2SiO4) bioceramics. *Ceram Inter* 33: 83–8.

Ni, S., and Chang, J., 2009. *In vitro* degradation, bioactivity, and cytocompatibility of calcium silicate, dimagnesium silicate, and tricalcium phosphate bioceramics. *J Biomater Appl* 24: 139–58.

Ni, S., Chang, J., and Chou, L., 2006. A novel bioactive porous CaSiO3 scaffold for bone tissue engineering. *J Biomed Mater Res A* 76: 196–205.

Ni, S., Chang, J., and Chou, L., 2008. *In vitro* studies of novel CaO-SiO2-MgO system composite bioceramics. *J Mater Sci Mater Med* 19: 359–67.

Ni, S., Lin, K., Chang, J., et al., 2008. Beta-CaSiO3/beta-Ca3(PO4)2 composite materi-als for hard tissue repair: *In vitro* studies. *J Biomed Mater Res A* 85: 72–82.

Nonami, T., and Tsutsumi, S., 1999. Study of diopside ceramics for biomaterials. *J Mater Sci Mater Med* 10: 475–9.

Ou, J., Kang, Y., Huang, Z., et al., 2008. Preparation and *in vitro* bioactivity of novel merwinite ceramic. *Biomed Mater* 3: 015015.

Peng, W., Liu, W., Zhai, W., et al., 2011. Effect of tricalcium silicate on the proliferation and odontogenic differentiation of human dental pulp cells. *J Endod* 37: 1240–6.

Ramaswamy, Y., Wu, C., Dunstan, C. R., et al., 2009. Sphene ceramics for orthopedic coating applications: An *in vitro* and *in vivo* study. *Acta Biomater* 5: 3192–204.

Ramaswamy, Y., Wu, C., Van Hummel, A., et al., 2008. The responses of osteoblasts, osteoclasts and endothelial cells to zirconium modified calcium-silicate-based ceramic. *Biomaterials* 29: 4392–402.

Ramaswamy, Y., Wu, C., Zhou, H., et al., 2008. Biological response of human bone cells to zinc-modified Ca-Si-based ceramics. *Acta Biomater* 4: 1487–97.

Schwarz, K., 1973. A bound form of silicon in glycosaminoglycans and polyuronides. *Proc Natl Acad Sci USA* 70: 1608–12.

Sun, H., Wu, C., Dai, K., et al., 2006. Proliferation and osteoblastic differentiation of human bone marrow-derived stromal cells on akermanite-bioactive ceramics. *Biomaterials* 27: 5651–7.

Tavangarian, F., and Emadi, R., 2011a. Effects of mechanical activation and chlorine ion on nanoparticle forsterite formation. *Mater Lett* 65: 126–9.

Tavangarian, F., and Emadi, R., 2011b. Improving degradation rate and apatite forma-tion ability of nanostructure forsterite. *Ceram Inter* 37: 2275–80.

Tavangarian, F., and Emadi, R., 2011c. Nanostructure effects on the bioactivity of for-sterite bioceramic. *Mater Lett* 65: 740–3.

Tavangarian, F., and Emadi, R., 2011d. Synthesis and characterization of spinel fors-terite nanocomposites. *Ceram Inter* 37: 2543–8.

Valerio, P., Pereira, M. M., Goes, A. M., et al., 2004. The effect of ionic products from bioactive glass dissolution on osteoblast proliferation and collagen production. *Biomaterials* 25: 2941–8.

Varlet, A., and Dauchy, P., 1983. Plaster of Paris pellets containing antibiotics in the treatment of bone infection. New combinations of plaster with antibiotics. *Rev Chir Orthop Reparatrice Appar Mot* 69: 239–44.

Wang, C., Xue, Y., Lin, K., et al., 2012. The enhancement of bone regeneration by a combination of osteoconductivity and osteostimulation using beta-CaSiO3/beta-Ca3(PO4)2 composite bioceramics. *Acta Biomater* 8: 350–60.

Wu, C., and Chang, J., 2004. Synthesis and apatite-formation ability of akermanite. *Mater Lett* 58: 2415–7.

Wu, C., and Chang, J., 2006. A novel akermanite bioceramic: Preparation and characteristics. *J Biomater Appl* 21: 119–29.

Wu, C., and Chang, J., 2007. Degradation, bioactivity, and cytocompatibility of diopside, akermanite, and bredigite ceramics. *J Biomed Mater Res B Appl Biomater* 83: 153–60.

Wu, C., and Chang, J., 2007. Synthesis and *in vitro* bioactivity of bredigite powders. *J Biomater Appl* 21: 251–63.

Wu, C., and Chang, J., Forthcoming. Silicate bioceramics for bone tissue regeneration. *J Inorg Mater.*

Wu, C., Chang, J., Ni, S., et al., 2006. *In vitro* bioactivity of akermanite ceramics. *J Biomed Mater Res A* 76: 73–80.

Wu, C., Chang, J., Wang, J., et al., 2005. Preparation and characteristics of a calcium magnesium silicate (bredigite) bioactive ceramic. *Biomaterials* 26: 2925–31.

Wu, C., Chang, J., and Zhai, W., 2005. A novel hardystonite bioceramic: Preparation and characteristics. *Ceram Inter* 31: 27–31.

Wu, C., Chang, J., Zhai, W., et al., 2006. Porous akermanite scaffolds for bone tissue engineering: Preparation, characterization, and *in vitro* studies. *J Biomed Mater Res B Appl Biomater* 78: 47–55.

Wu, C., Chang, J., Zhai, W., et al., 2007. A novel bioactive porous bredigite (Ca$_{(7)}$MgSi$_{(4)}$O$_{(16)}$) scaffold with biomimetic apatite layer for bone tissue engineering. *J Mater Sci Mater Med* 18: 857–64.

Wu, C., Fan, W., Zhou, Y. H., et al., 2012. 3D-printing of highly uniform CaSiO3 ceramic scaffolds: Preparation, characterization and *in vivo* osteogenesis. *J Mater Chem* 22: 12288–95.

Wu, C., Ramaswamy, Y., and Zreiqat, H., 2010. Porous diopside (CaMgSi(2)O(6)) scaffold: A promising bioactive material for bone tissue engineering. *Acta Biomater* 6: 2237–45.

Wu, C., Ramaswamy, Y., Boughton, P., et al., 2008. Improvement of mechanical and biological properties of porous CaSiO3 scaffolds by poly(D,L-lactic acid) modification. *Acta Biomater* 4: 343–53.

Wu, C., Ramaswamy, Y., Chang, J., et al., 2008. The effect of Zn contents on phase composition, chemical stability and cellular bioactivity in Zn-Ca-Si system ceramics. *J Biomed Mater Res B Appl Biomater* 87: 346–53.

Wu, C., Ramaswamy, Y., Gale, D., et al., 2008. Novel sphene coatings on Ti-6Al-4V for orthopedic implants using sol-gel method. *Acta Biomater* 4: 569–76.

Wu, C., Ramaswamy, Y., Kwik, D., et al., 2007. The effect of strontium incorporation into CaSiO3 ceramics on their physical and biological properties. *Biomaterials* 28: 3171–81.

Wu, C., Ramaswamy, Y., Liu, X., et al., 2009. Plasma-sprayed CaTiSiO5 ceramic coating on Ti-6Al-4V with excellent bonding strength, stability and cellular bioactivity. *J R Soc Interface* 6: 159–68.

Wu, C., Ramaswamy, Y., Soeparto, A., et al., 2008. Incorporation of titanium into calcium silicate improved their chemical stability and biological properties. *J Biomed Mater Res A* 86: 402–10.

Wu, C., and Zreiqat, H., 2010. Porous bioactive diopside (CaMgSi(2)O(6)) ceramic microspheres for drug delivery. *Acta Biomater* 6: 820–9.

Xia, L., Zhang, Z., Chen, L., et al., 2011. Proliferation and osteogenic differentiation of human periodontal ligament cells on akermanite and beta-TCP bioceramics. *Eur Cell Mater* 22: 68–82.

Xu, S., Lin, K., Wang, Z., et al., 2008. Reconstruction of calvarial defect of rabbits using porous calcium silicate bioactive ceramics. *Biomaterials* 29: 2588–96.

Xynos, I. D., Edgar, A. J., Buttery, L. D., et al., 2000. Ionic products of bioactive glass dissolution increase proliferation of human osteoblasts and induce insulin-like growth factor II mRNA expression and protein synthesis. *Biochem Biophys Res Commun* 276: 461–5.

Zhai, W., Lu, H., Chen, L., et al., 2012. Silicate bioceramics induce angiogenesis during bone regeneration. *Acta Biomater* 8: 341–9.

Zhang, M. L., and Chang, J., 2012. Preparation and characterization of Sr-hardystonite (Sr2ZnSi2O7) for bone repair applications. *Mater Sci Eng C* 32: 184–8.

Zhang, M., Zhai, W., and Chang, J., 2010. Preparation and characterization of a novel willemite bioceramic. *J Mater Sci Mater Med* 21: 1169–73.

Zhang, M., Zhai, W., Lin, K., et al., 2010. Synthesis, *in vitro* hydroxyapatite forming ability, and cytocompatibility of strontium silicate powders. *J Biomed Mater Res B Appl Biomater* 93: 252–7.

Zhang, W., Shen, Y., Pan, H., et al., 2011. Effects of strontium in modified biomaterials. *Acta Biomater* 7: 800–8.

Zhao, W., and Chang, J., 2005. Preparation and characterization of novel tricalcium silicate bioceramics. *J Biomed Mater Res A* 73: 86–9.

Zhao, W., Chang, J., and Zhai, W., 2008. Self-setting properties and *in vitro* bioactivity of Ca3SiO5/CaSO4.1/2H2O composite cement. *J Biomed Mater Res A* 85: 336–44.

Zhao, W., Chang, J., Wang, J., et al., 2007. *In vitro* bioactivity of novel tricalcium silicate ceramics. *J Mater Sci Mater Med* 18: 917–23.

Zhao, W., Wang, J., Zhai, W., et al., 2005. The self-setting properties and *in vitro* bioactivity of tricalcium silicate. *Biomaterials* 26: 6113–21.

Zhao, Y. K., Ning, C. Q., and Chang, J., 2009. Sol-gel synthesis of Na(2)CaSiO(4) and its *in vitro* biological behaviors. *J Sol-Gel Sci Tech* 52: 69–74.

Zhong, H. B., Wang, L. J., Fan, Y. C., et al., 2011. Mechanical properties and bioactivity of beta-Ca(2)SiO(4) ceramics synthesized by spark plasma sintering. *Ceram Inter* 37: 2459–65.

Zhou, Y., Wu, C., and Xiao, Y., 2012. The stimulation of proliferation and differentiation of periodontal ligament cells by the ionic products from Ca(7)Si(2)P(2) O(16) bioceramics. *Acta Biomater* 8: 2307–16.

Zreiqat, H., Ramaswamy, Y., Wu, C., et al., 2010. The incorporation of strontium and zinc into a calcium-silicon ceramic for bone tissue engineering. *Biomaterials* 31: 3175–84.

3

Functional Mesoporous Silica Nanoparticles with a Core-Shell Structure for Controllable Drug Delivery

Yufang Zhu

CONTENTS

3.1 Introduction

3.1.1 Amorphous Silica Nanoparticles

Silica (SiO_2) is the most abundant component of the earth's crust except carbon, and it can be divided into crystalline and amorphous forms. Crystalline silica is known to cause adverse health effects such as silicosis, whereas amorphous silica is considered to be biocompatible and nontoxic (Park et al. 2011; Uboldi et al. 2012). Hence, amorphous silica, especially in the form of nanoparticles, has been employed in a wide range of industrial and biomedical applications including cosmetics, food additives, and drug delivery systems (Hirsch et al. 2003; Huang et al. 2005; Dekkers et al. 2010; Shi et al. 2010).

In general, amorphous silica is highly inert and stable in comparison with organic systems such as liposomes, dendrimers, and polymers, which can be suitable carriers for a wide range of drug delivery systems in the physiological environment (Barbé et al. 2004). On the other hand, amorphous silica nanoparticles are degradable and undergo hydrolysis to form silicic acid, $Si(OH)_4$, in aqueous solution. *In vivo*, silicic acid from the degradation of silica nanoparticles can diffuse through the bloodstream or lymph, and is excreted through urine (Finnie et al. 2009). Furthermore, for the application in drug delivery, silica nanoparticles can be surface functionalized with adequate organic reagents due to the abundant reactive silanol groups (-OH) on the surface and, by doing so, an active layer for the absorption of desired drug molecules could be generally produced.

Currently, the simplest way to synthesize amorphous silica nanoparticles is the Stöber method based on the sol-gel technique (Stöber et al. 1968). Using this procedure, amorphous silica nanoparticles with a large variety of sizes can be synthesized by controlling the synthetic conditions. However, the pore volume and surface area of these nanoparticles are usually lower, which limits the drug adsorption capacity on silica nanoparticles. Therefore, amorphous silica nanoparticles with high surface area and pore volume are pursued for high-efficient drug delivery.

3.1.2 Mesoporous Silica Nanoparticles

Ordered mesoporous silica materials, a new family of molecular sieves called M41S, were discovered by Mobil researchers in 1992 (Kresge et al. 1992). They are synthesized in the presence of assembled surfactant micelle templates, which serve as structure-directing agents for polymerizing silica component by electrostatic interaction. The most well-known and common mesoporous silica includes MCM-41, MCM-48, and SBA-15. Mesoporous silica materials are amorphous and have unique properties including high surface area, large pore volume, uniform and tunable pore size, nontoxic nature, well-defined surface properties for functionalization, and good biocompatibility (He and Shi 2011). On the one hand, textural properties of mesoporous silica (such as high surface area, large pore volume, and mesoporous channels with tunable pore size) provide the possibility to load a high amount of drugs within mesoporous silica carriers (Muñoz et al. 2003). On the other hand, there are abundant silanol groups on the surfaces of mesoporous channels and the outer surfaces of mesoporous silica nanoparticles, which facilitate the surface functionalization to allow for a better control over the drug diffusion kinetics (Tasciotti et al. 2008; Van Speybroeck et al. 2009). In addition, functional materials, such as magnetic nanoparticles, luminescent materials, and polymers, can be combined with mesoporous silica to form functional mesoporous silica composites, which induce mesoporous silica composites as multifunctional platforms to realize the targeted controlled drug delivery or imaging (Yang et al. 2012). Therefore, since the report on MCM-41 mesoporous silica for drug delivery by Vallet-Regí and colleagues in 2001, mesoporous silica nanoparticles have been considered to be excellent candidates as carriers for drug delivery (Manzano and Vallet- Regí 2010; Vivero-Escoto et al. 2010; Tang et al. 2012).

Recently, great endeavors have been made in the structure design and functional optimization to advance the development of mesoporous silica-based drug-delivery systems (Li et al. 2012). Meanwhile, the biological effects of mesoporous silica nanoparticles at different levels from molecule, cell, and blood to organ/tissue, involving cytotoxicity, biodegradability, blood compatibility, biodistribution, and excretion have also become current hot research topics in the field of mesoporous silica nanoparticles for biomedicine (He, Zhang, et al. 2009; He, Zhang, Shi, et al. 2010; He, Shi, et al. 2010; He et al. 2011). To date, various functional mesoporous silica composites with different structures have been developed, such as embedding of functional materials in mesoporous channels, the core-shell structure, and functionalization on the surface of mesoporous silica (Lian et al. 2009; Chang et al. 2010; Rosenholm, Zhang, et al. 2011). Actually, design of different structured functional mesoporous silica composites is based on their applications. It has been demonstrated that the core-shell structured functional mesoporous silica nanoparticles can combine new functions in the platform without blocking the mesoporous channels. Therefore, the core-shell structured functional mesoporous silica nanoparticles are desirable for drug delivery and has been one of the most active research topics on the applications of mesoporous silica nanoparticles. In this chapter, we will highlight the recent studies on functional mesoporous silica nanoparticles with a core-shell structure for controllable drug delivery.

3.2 Functional Hollow Mesoporous Silica Nanoparticles for Controllable Drug Delivery

It has been accepted that mesoporous silica nanoparticles are one of the most excellent candidates as carriers for drug delivery (Manzano and Vallet- Regí 2010; Vivero-Escoto et al. 2010; Tang et al. 2012). Compared to conventional mesoporous silica nanoparticles, hollow mesoporous silica (HMS) nanoparticles, also called silica nanoparticles with hollow core-mesoporous shell structure, possess much higher drug-loading capacity due to the hollow cores providing more space to load drug molecules. On the other hand, the penetrating mesopores in the shells guarantee the facile transport of drug molecules into and out of the hollow cores (Zhu, Shi, Chen, et al. 2005; Zhu, Shi, Li, et al. 2005; Zhu, Shi, Shen, Chen, et al. 2005; Zhu, Shi, Shen, Dong, et al. 2005; Zhao et al. 2009). Therefore, HMS nanoparticles have attracted great and increasing attention for drug delivery.

3.2.1 Preparation of Hollow Mesoporous Silica Nanoparticles

To date, great efforts have been devoted to the preparation of HMS nanoparticles with controllable particle and pore sizes (Li et al. 2003; Tan and Rankin 2005; Zhu, Shi, Chen, et al. 2005; Zhu, Shi, Shen, Dong, et al. 2005; Zhao et al. 2009; Chen, Chen, Guo, et al. 2010; Du et al. 2011; Fang et al. 2011; Zhu et al. 2011; Lim et al. 2012). Traditional preparation methodologies of HMS nanoparticles are so-called soft-/hard-templating routes (Li et al. 2003; Tan and Rankin 2005; Zhu, Shi, Chen, et al. 2005; Zhu, Shi, Shen, Dong, et al. 2005; Zhao et al. 2009; Du et al. 2011; Zhu et al. 2011; Lim et al. 2012) including the fabrication of uniform soft/hard templates and deposition of the mesoporous silica shells. The soft templates could be emulsion drops or micellar aggregates. The hard templates could be Fe_2O_3 nanoparticles, polystyrene spheres, or carbon spheres. The hollow cores could be obtained by removing the templates via calcination or dissolution using suitable reagents or solvents. In general, the structural parameters of HMS nanoparticles via the soft-/hard-templating routes can be regulated by the adjustment of synthetic conditions (e.g., reaction time, temperature, precursor concentration, etc.) to endow materials with desired structures and properties. Zhu, Shi, Chen, et al. (2005) reported for the first time a facile route to HMS spheres with penetrating pore channels across the shells by an approach using PVP aggregates and cetyltrimethylammonium bromide (CTAB) as cotemplates. Here, PVP aggregates and CTAB surfactant serve as the templates for the cores and mesoporous structure, respectively (see Figure 3.1). Later, they also synthesized HMS nanoparticles with particle diameter of ca. 100 nm using the colloidal carbon spheres as templates (Zhu, Shi, Shen, et al. 2005). The particle size and shell thickness of HMS nanoparticles can be turned through changing the carbon spheres and the addition of a silica source.

Recently, many efforts have also been made to develop new strategies for the preparation of HMS nanoparticles (Chen, Chen, Guo, et al. 2010; Fang et al. 2011). Chen, Chen, Guo, et al. (2010) have developed a simple synthetic strategy, namely, the structure difference-based selective etching process, to prepare HMS nanoparticles, in which the silica core was selectively etched away from the silica core/mesoporous silica shell structure while the mesoporous silica shell was kept almost intact (see Figure 3.2). For this silica core-shell template, the composition between core and shell is the same, while the structures are different. This leads to different dissolution behaviors of the core and shell in Na_2CO_3 solution (Route A) and in ammonia solution under

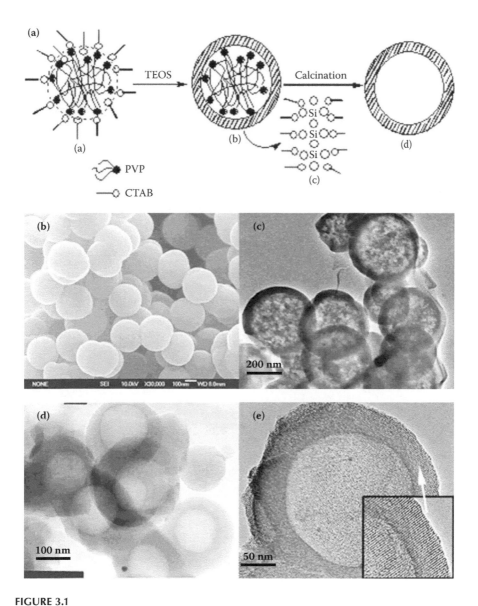

FIGURE 3.1
(a) Schematic drawing of the formation process of HMS nanoparticles. (b) SEM image of HMS nanoparticles. (c) TEM image of the as-synthesized HMS nanoparticles. (d) TEM and (e) HRTEM images of the calcined HMS nanoparticles. (Reprinted from Zhu Y., Shi J., Chen H., et al., *Micropor. Mesopor. Mater.* 84: 218–222, 2005, with permission from Elsevier Ltd.)

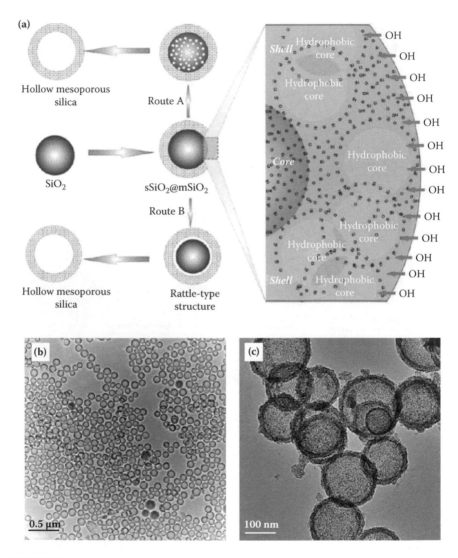

FIGURE 3.2
(a) Schematics of the formation of hollow/rattle-type mesoporous silica spheres. Route A represents the selective-etching procedure in Na$_2$CO$_3$ solution and Route B for that in ammonia solution under hydrothermal treatment. (b) TEM image of HMS nanoparticles obtained by treating sSiO$_2$@mSiO$_2$ in 0.6 M Na$_2$CO$_3$ solution at 80°C for 0.5 h. (c) HMS nanoparticles obtained by treating sSiO2@mSiO2 in 0.24 M ammonia solution at 150°C for 24 h. (Reprinted with permission from Chen Y., Chen H., Guo L., et al. *ACS Nano* 4: 529–539, 2010, Copyright 2009, American Chemical Society.)

hydrothermal treatment (Route B). Thus, monodisperse HMS nanoparticles with tunable particle/pore sizes can be obtained by selectively removing the solid silica cores (Figure 3.2) (Chen, Chen, Guo, et al. 2010).

3.2.2 Functional Hollow Mesoporous Silica Nanoparticles for Drug Delivery

Many studies have demonstrated that HMS nanoparticles exhibit higher drug-loading capacity compared to conventional mesoporous silica nanoparticles. For example, Zhao et al. (2009) have reported that the maximum ibuprofen (IBU) loading capacity of HMS nanoparticles (726 mg/g) is much higher than the reported maximum IBU loading capacity of MCM-41 (358 mg/g), although the surface area and pore volume of HMS nanoparticles (455 m^2/g, 0.59 cm^3/g) are much lower than those of MCM-41 (1152 m^2/g, 0.99 cm^3/g). However, pure HMS nanoparticles as carriers for drug delivery only show the sustained drug release behavior and are difficult to control the drug release rate (the pore diameter and pore structure type determine the drug release rate) (Vallet-Regí et al. 2001; Andersson et al. 2004; Izquierdo-Barba et al. 2009).

Recently, many efforts have been made to improve the drug delivery properties by functionalizing HMS nanoparticles (Zhu, Shi, Li, et al. 2005; Zhu and Shi 2007; Zhang et al. 2010). On the one hand, for drug molecules loading in HMS nanoparticles, some drug molecules are located in the hollow cores, and others are loaded in the mesoporous channels and on the surfaces of mesoporous walls through weak interactions, such as hydrogen bonding, physical adsorption, and electrostatic interactions. Therefore, functionalization of the mesoporous shells with appropriate groups, presenting attractive interactions with drug molecules and slightly turning the pore diameter, may provide an effective control on the drug loading and release rate. Zhu, Shi, Li, et al. (2005) functionalized hollow mesoporous silica spheres with cubic pore network (HMSC) with 3-aminopropyltriethoxysilane (N-TES), 3-(2-aminoethylamino)- propyltrimethoxysilane (NN-TES), and (3-trimethoxysilylpropyl)diethylenetriamine (NNN-TES) at different levels by a simple one-step method and postmodification process. With the increase of the amount of functional groups introduced, the IBU drug release rate becomes lower. At the same amount of functional groups, the IBU drug release rate follows the order: NNN-HMSC < NN-HMSC < NHMSC. Obviously, functionalization of HMS nanoparticles with appropriate groups can control the drug release rate but still exhibit the sustained release behavior.

On the other hand, compared to the sustained release system, the stimuli responsive controlled release system can achieve a site selective controlled release pattern, which can improve the therapeutic efficacy. Zhu and Shi (2007) proposed a strategy to functionalize HMS nanoparticles with the pH-responsive polyelectrolyte multilayers [the polyelectrolyte pair, sodium poly(styrene sulfonate) (PSS), and polycation poly(allylamine hydrochloride)

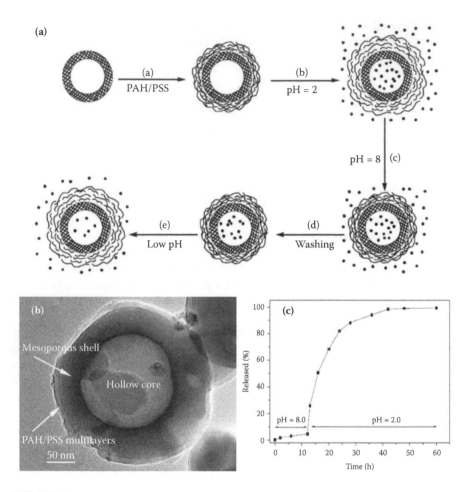

FIGURE 3.3
(a) Scheme of polyelectrolyte multilayers (PEM) (consisting of PSS and PAH layers) coating on
HMS nanoparticles and pH-controlled loading and release of drug molecules. (b) TEM image
of PEM/HMS nanoparticles. (c) pH-controlled release switching of gentamicin molecules from
PEM/HMS system. (Reprinted from Zhu Y. and Shi J., *Micropor. Mesopor. Mater.* 103: 243–249,
2007, with permission from Elsevier Ltd.)

(PAH)] for controlling the drug release behavior. As shown in Figure 3.3,
HMS nanoparticles are coated by a layer-by-layer (LbL) technique with
oppositely charged PAH and PSS alternatively. When this system is put in a
medium of low pH value, the PAH/PSS multilayers will open the pathways
for drug molecule diffusion into HMS nanoparticles. As the pH value of the
medium increases, PAH/PSS multilayers will close the pathways, and drug
molecules are loaded in HMS nanoparticles. To obtain the controlled drug
release, one simply puts the system loaded with drug molecules in a medium
of low pH value. In this way, the system not only has high drug-loading

capacity and stable structure but also possesses a convenient pH-controlled loading and release property.

3.3 Functional Mesoporous Silica Nanoparticles with a Core-Shell Structure for Controllable Drug Delivery

Mesoporous silica nanoparticles have been intensively studied as drug delivery carriers due to their unique structure and surface properties. However, pure mesoporous silica suffers from some limitations in many applications (He and Shi 2011). For example, pure mesoporous silica cannot realize the targeted drug delivery, and cannot track or evaluate the efficiency of drug release in disease diagnosis and therapy. Therefore, the combination with mesoporous silica and functional materials to form the core-shell structure is a smart strategy to solve the aforementioned limitations, especially to form magnetic or luminescent mesoporous silica nanoparticles. Thus, these mesoporous silica nanoparticles with a core-shell structure can be used as multifunctional platforms for simultaneous targeted drug delivery, fast diagnosis, and efficient therapy. In this part, we will introduce magnetic, luminescent, and other multifunctional mesoporous silica nanoparticles with a core-shell structure for controllable drug delivery.

3.3.1 Magnetic Mesoporous Silica Nanoparticles for Controllable Drug Delivery

Currently, magnetic nanoparticles, an important class of inorganic materials, are especially attractive for targeted drug delivery and hyperthermia application (Kumar and Mohammad 2011; Laurent et al. 2011). Magnetic targeting provides the ability to guide the drug delivery systems to the desired location by means of an external magnetic field and keep them until the therapy is complete, which will facilitate the therapeutic efficiency and reduce the side effect of the toxic drugs before targeting the desired positions (Laurent et al. 2011). On the other hand, hyperthermia using superparamagnetic nanoparticles under alternating magnetic fields has been an efficient strategy in cancer therapy due to the heating ability of superparamagnetic nanoparticles, as a result of Brownian rotation and Néel relaxation mechanisms (Kumar and Mohammad 2011). Among magnetic materials, γ-Fe_2O_3 and Fe_3O_4 have been most intensively studied. However, pure iron oxide is prone to aggregation because of anisotropic dipolar attraction and rapid biodegradation when they are exposed to biological systems directly (Ruíz-Hernández et al. 2007; Zhou et al. 2007). Furthermore, pure iron oxide nanoparticles as drug carriers have difficulty anchoring drug molecules and possess relative lower drug loading capacity. Therefore, the combination of iron oxide nanoparticles and mesoporous silica to form magnetic mesoporous silica nanoparticles would overcome the limitations and represent a significant advance in the field of drug delivery (Liong et al. 2008; Zhang et al. 2008; Huang et al. 2012; Liu et al. 2012).

Recently, preparation of magnetic mesoporous silica nanoparticles has received much interest. Different types of magnetic mesoporous silica nanoparticles have been reported for controllable drug delivery, including (1) embedding iron oxide nanocrystals into mesoporous silica matrices (Yiu et al. 2010); (2) the core-shell structured nanoparticles with iron oxide core and mesoporous silica shell (Wu et al. 2004; Zhao et al. 2005; Kim et al. 2006; Deng et al. 2008; Kim et al. 2008; Zhao et al. 2008; Lin et al. 2009; Lin and Haynes 2009; Zhu et al. 2009; Chen, Chen, Zeng, et al. 2010; Fu et al. 2010; Liu et al. 2010; Tan et al. 2010; Zhu, Fang, and Kaskel 2010; Zhu, Ikoma, et al. 2010; Chen, Chen, Ma, et al. 2011; Gai et al. 2011; Rosenholm, Sahlgren, et al. 2011; Wu et al. 2011; Zhang et al. 2011; Zhu, Jian, and Wang 2011; Zhu, Meng, Gao, et al. 2011; Zhu, Meng, and Hanagata 2011); and (3) capping the pores of mesoporous silica with iron oxide nanocrystals for stimuli-responsive controlled drug release (Giri et al. 2005; Gan et al. 2011). Among these magnetic mesoporous silica nanoparticles, the core-shell structured nanoparticles with iron oxide core and mesoporous silica shell are considered as the most important and desirable structure for drug delivery combined with targeting or hyperthermia functions.

For the core-shell structure, the most straightforward synthetic strategy is the so-called bottom-up approach, where the core and shell are prepared in an inside-to-outside order. In 2004, Wu et al. for the first time reported magnetic mesoporous silica particles with Fe_3O_4 core and mesoporous silica shell, which are prepared via deposition of mesoporous silica film on magnetic Fe_3O_4 cores by using cetyltrimethylammonium chloride (CTACl) micelles as structure-directing agents. Since then, different approaches have been developed to prepare magnetic mesoporous silica nanoparticles with a core-shell structure. Also, these magnetic mesoporous silica nanoparticles have been used as carriers for application in drug delivery.

Self-assembly, phase transfer, and microemulsion techniques are effective routes to magnetic mesoporous silica nanoparticles with a core-shell structure. For hydrophobic ligand-capped magnetic nanocrystals, they can be successfully encapsulated into the mesoporous silica nanoparticles by using CTAB surfactant as a structure agent and a coagent for transferring the magnetic nanoparticles from oil to water, and the resulting nanoparticles can be formed by encapsulating one to several magnetic nanocrystals into one mesoporous silica nanoparticle. The size of these magnetic mesoporous silica nanoparticles can be easily controlled. Kim et al. (2006) reported the synthesis of monodisperse and size-controllable core-shell mesoporous silica nanoparticles by using single Fe_3O_4 nanocrystals as cores (designated as Fe_3O_4@mSiO$_2$) (see Figure 3.4). In this approach, hydrophobic ligand-capped

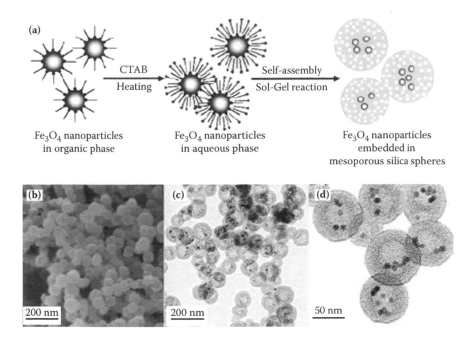

FIGURE 3.4
(a) Synthetic procedure of monodisperse and size-controllable core-shell $Fe_3O_4@mSiO_2$ nanoparticles. (b) FE-SEM. (c) TEM and (d) HRTEM images of $Fe_3O_4@mSiO_2$ nanoparticles. (Reprinted with permission from Kim J., Lee J. E., Lee J., et al., *J. Am. Chem. Soc.* 128: 688–699, Copyright 2006, American Chemical Society.)

magnetic Fe_3O_4 nanocrystals were first transferred from the organic phase to the aqueous phase using CTAB surfactant. The subsequent sol-gel reaction of TEOS in the aqueous solution containing CTAB and oleic acid-stabilized magnetic nanocrystals resulted in magnetic mesoporous silica nanoparticles with a core-shell structure. Here, CTAB serves not only as the stabilizing surfactant for the transfer of hydrophobic Fe_3O_4 nanocrystals to the aqueous phase but also as the organic template for the formation of mesopores in the sol-gel reaction. The $Fe_3O_4@mSiO_2$ particles can be collected after the removal of the CTAB templates from the as-synthesized nanoparticles by extraction in acidic ethanol solution (pH 1.4). The same research group also reported the synthesis of magnetic fluorescent delivery vehicles by encapsulating monodisperse magnetic and semiconductor nanocrystals in uniform mesoporous silica nanoparticles (Kim et al. 2008). Similarly, Zhao et al. (2005) synthesized magnetic silica nanoparticles with highly ordered mesostructures by using the phase transfer method. For this approach, the resulting magnetic mesoporous silica nanoparticles possess a uniform nanosize ($\approx$90–140 nm) and a highly ordered mesostructure, and, more important, the pore size can be changed from 2.4 to 3.4 nm and the saturation magnetization values can be changed from 0.43 to 7.62 emu/g (Zhao et al. 2005).

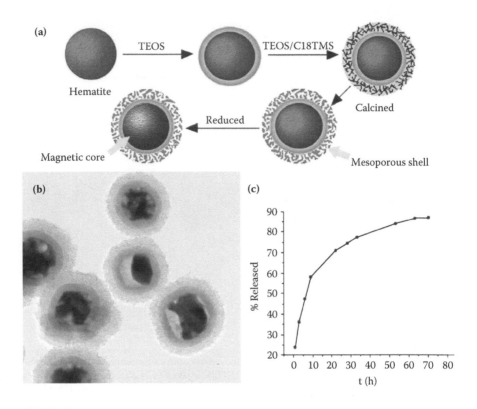

FIGURE 3.5
(a) Illustration of synthesis of Fe_3O_4/Fe@nSiO$_2$@mSiO$_2$ nanoparticles. (b) TEM image of Fe_3O_4/Fe@nSiO$_2$@mSiO$_2$ nanoparticles. (c) IBU release profile from Fe_3O_4/Fe@nSiO$_2$@mSiO$_2$ nanoparticles. (Reprinted with permission from Zhao W., Gu J., Zhang L., et al., *J. Am. Chem. Soc.* 127: 8916–8917, Copyright 2005, American Chemical Society.)

However, the aforementioned techniques are limited to synthesize magnetic mesoporous silica nanoparticles with poor magnetic response due to the difficulty in increasing the amount of magnetic nanocrystals in the whole nanoparticles (Kim et al. 2006, 2008; Lin and Haynes 2009). Recently, many efforts have been made to synthesize magnetic silica nanoparticles with large magnetic nanoparticle cores ($\approx$100–200 nm) and mesoporous silica shells in order to increase the mass fraction of magnetic nanocrystals (Zhao et al. 2005; Deng et al. 2008; Fu et al. 2010; Gai et al. 2011; Rosenholm, Sahlgren, et al. 2011). Zhao et al. (2005) reported a novel strategy to fabricate uniform magnetic nanoparticles with a magnetic core/mesoporous silica shell structure (see Figure 3.5). Herein, uniform α-Fe$_2$O$_3$ nanoparticles were employed as the initial cores, a thin and dense silica layer was deposited on the surface of α-Fe$_2$O$_3$ nanoparticles in order to protect α-Fe$_2$O$_3$ core from leaching into the mother system under acidic circumstances. Then, mesoporous silica shell was formed from simultaneous sol-gel polymerization

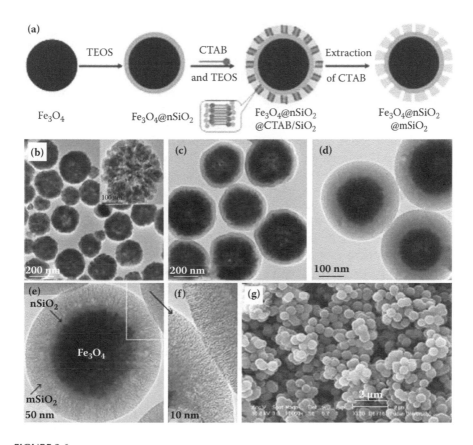

FIGURE 3.6
(a) The formation of $Fe_3O_4@nSiO_2@mSiO_2$ nanoparticles. TEM images of (b) Fe_3O_4 particles, (c) $Fe_3O_4@nSiO_2$, (d–f) $Fe_3O_4@nSiO_2@mSiO_2$ nanoparticles, and (g) SEM image of $Fe_3O_4@nSiO_2@$ $mSiO_2$ nanoparticles. (Reprinted with permission from Deng Y., Qi D., Deng C., et al., *J. Am. Chem. Soc.* 130: 28–29, Copyright 2008, American Chemical Society.)

of TEOS and *n*-octadecyltrimethoxysilane (C18TMS) followed by the removal of the organic group. Finally, α-Fe_2O_3 cores of the nanoparticles were reduced in a flowing H_2/N_2 gas mixture to produce the final magnetic mesoporous silica nanoparticles with a core-shell structure ($Fe_3O_4/$ $Fe@nSiO_2@mSiO_2$ nanoparticles). Using IBU as a model drug to explore the capability as drug carriers, the loading amount of IBU was ca. 12 wt%, as assessed by TG analysis, and the release in a simulated body fluid (SBF) can keep over 70 h, allowing for the potential application in targeting drug delivery. The prepared magnetic mesoporous silica nanoparticles have a high magnetization value (27.3 emu/g), but its ferromagnetic property may limit its practical application in some areas. Deng et al. (2008) synthesized superparamagnetic particles with Fe_3O_4 cores and perpendicularly aligned mesoporous silica shells through a three-step process (see Figure 3.6). First, nonporous silica/Fe_3O_4 nanoparticles were prepared by coating a thin silica layer on uniform Fe_3O_4 particles. Second, through a surfactant templating approach with CTAB as the template, mesoporous silica shells were deposited on the nonporous silica/Fe_3O_4 nanoparticles. Third, the CTAB templates were removed by extraction to form perpendicularly aligned mesoporous silica shells with thickness of around 70 nm. Gai et al. (2011) designed novel fibrous-structured magnetic mesoporous silica particles with a core-shell structure (denoted as Fe_3O_4/FMSMs). The Fe_3O_4/FMSMs exhibit a sustained drug release profile, sufficient magnetic responsibility, and redispersibility to the external magnetic field. Also, the Fe_3O_4/FMSMs nanoparticles could be used as a therapeutically effective intracellular drug delivery system for doxorubicin (DOX) delivery.

For the aforementioned magnetic mesoporous silica nanoparticles with a core-shell structure, the magnetic cores are compactly stuck to the silica shells, and therefore their magnetic properties may be compromised to some extent. Magnetic mesoporous silica nanoparticles with rattle-type or yolk/ shell structure can prove to be a solution to this problem (Zhao et al. 2008; Lin et al. 2009; Zhu et al. 2009; Chen, Chen, Zeng, et al. 2010; Liu et al. 2010; Tan et al. 2010; Zhu, Fang, and Kaskel 2010; Zhu, Ikoma, et al. 2010; Chen, Chen, Ma, et al. 2011; Wu et al. 2011; Zhang et al. 2011; Zhu, Jian, and Wang 2011; Zhu, Meng, Gao, et al. 2011; Zhu, Meng, and Hanagata 2011), as such nanoparticles possess unique structures with interstitial spaces between mesoporous silica shells and magnetic cores, that is, movable magnetic nanoparticles encapsulated in hollow mesoporous silica nanoparticles. Also, rattle-type magnetic mesoporous silica nanoparticles are designed and applied as controlled drug delivery carriers, because the large cavity volume between yolk and shell can provide more space for loading drug molecules. Therefore, rattle-type magnetic mesoporous silica nanoparticles can be used as multifunctional platforms for biomedical applications, such as diagnostic agent and targeted drug delivery. Shi's group has prepared a series of rattle-type Fe_3O_4@ $mSiO_2$ mesoporous ellipsoids/spheres with a single or double mesoporous silica shells by using the so-called etching approach (Zhao et al. 2008; Chen, Chen, Guo, et al. 2010; Chen, Chen, Zeng, et al. 2010; Chen, Chen, Ma, et al. 2011; Wu et al. 2011). For example, Wu et al. (2011) first prepared Fe_2O_3/ SiO_2/$mSiO_2$ core-shell structures, and rattle-type Fe_3O_4@$mSiO_2$ mesoporous spheres were obtained by hydrothermally treating the Fe_2O_3/SiO_2/$mSiO_2$ core-shell structures and following the reduction in an H_2/N_2 atmosphere. They proved that the saturation magnetization value of rattle-type Fe_3O_4@ $mSiO_2$ mesoporous spheres (35.7 emu/g) is evidently higher than that of the corresponding core-shell structure with an intact middle silica layer (28.8 emu/g) due to the removal of the in-between silica layer (Wu et al. 2011). In addition, surface area and pore volume of the rattle-type Fe_3O_4@$mSiO_2$ mesoporous spheres are calculated to be 435 m^2/g and 0.58 cm^3/g, respectively, which are significantly higher than the sample with an intact middle silica layer (274 m^2/g and 0.38 cm^3/g). Furthermore, these rattle-type Fe_3O_4@$mSiO_2$ mesoporous spheres possess high efficiency docetaxel (DOC) loading ability

(180 mg DOC/g PEG/FA-Fe_3O_4@mSiO_2 nanoparticles) and excellent blood compatibility. Most important, the DOC-loaded Fe_3O_4@mSiO_2 nanoparticles showed the greater cytotoxicity than free DOX to induce MCF-7 cell death due to the drug release in cells. Therefore, compared with the normal core-shell magnetic mesoporous silica nanoparticles, rattle-type magnetic meso-porous silica nanoparticles are more suitable as excellent anticancer drug carriers for targeting drug release.

Zhu et al. (2009; Zhu, Ikoma, et al. 2010; Zhu, Jian, and Wang 2011) have also conducted extensive investigations into the preparation of rattle-type magnetic mesoporous silica nanoparticles with large cavities by using col-loidal carbon spheres as the templates. As shown in Figure 3.7, the first step involved the preparation of the colloidal carbon spheres adsorbed with iron precursor by a one-pot hydrothermal treatment. In the next step, the organosilicate-incorporated silica shells were deposited on the colloidal carbon spheres through the simultaneous sol-gel polymerization of TEOS and C18TMS. Finally, rattle-type Fe_3O_4@SiO_2 hollow mesoporous spheres were obtained after the calcination to remove the carbon templates and the organic groups of C18TMS, and then the reduction under H_2/Ar atmosphere (Zhu, Ikoma, et al. 2010). The particle sizes, mesoporous shell thicknesses, and Fe_3O_4 amounts can be readily controlled by adjusting the experimen-tal conditions. It has been studied that these Fe_3O_4@SiO_2 hollow mesopo-rous spheres had no cytotoxicity against Hela cells and exhibited relatively fast cell uptake localized near the nucleus. DOX released from Fe_3O_4@SiO_2 hollow mesoporous spheres had a sustained release pattern, and the DOX-loaded spheres exhibited greater cytotoxicity than free DOX (Zhu, Ikoma, et al. 2010).

Furthermore, these rattle-type Fe_3O_4@SiO_2 hollow mesoporous spheres can be modified with functional ligands or groups for multifunctional applications (Zhu, Fang, and Kaskel 2010; Zhu, Meng, Gao, et al. 2011; Zhu, Meng, and Hanagata 2011). Folate-conjugated Fe_3O_4@SiO_2 hollow mesopo-rous spheres have been developed as anticancer drug carriers for the tar-geted drug delivery system (Zhu, Fang, and Kaskel 2010). It is interesting that the drug delivery system combines the abilities of receptor-mediated and magnetic targeting, and the DOX-loaded Fe_3O_4@SiO_2-FA spheres exhibited greater cytotoxicity than free DOX and DOX-loaded Fe_3O_4@SiO_2 spheres due to the increase of cell uptake of anticancer drug delivery system mediated by the FA receptor. On the other hand, Zhu, Meng, Gao, et al. (2011) have also designed and constructed an enzyme-responsive carrier for co-delivery of drugs and genes using rattle-type Fe_3O_4@SiO_2 hollow mesoporous spheres (HMS)/poly(L-lysine) (PLL) core-shell particles (Figure 3.8). The first step involves the loading of fluorescein, a model drug, into HMS particles and the modification with 3-aminopropyltriethoxysilane on the surface of the fluorescein-loaded particles to obtain the modified fluorescein-loaded HMS (MFHMS) particles. In the next step, the MFHMS particles are coated by a layer-by-layer (LbL) assembly with negatively charged CpG ODN, a model

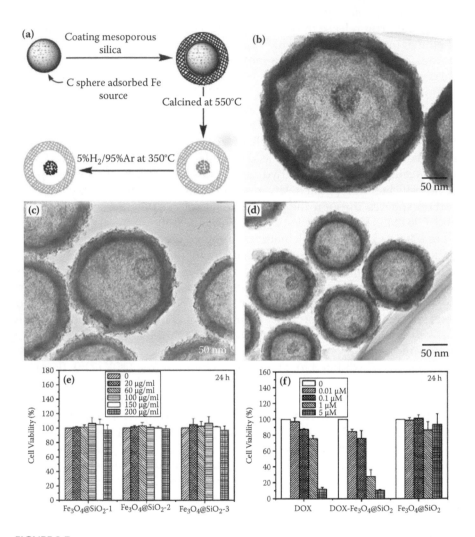

FIGURE 3.7

(a) Schematic procedure for the preparation of rattle-type $Fe_3O_4@SiO_2$ hollow mesoporous nanoparticles. (b–d) TEM images of $Fe_3O_4@SiO_2$ hollow mesoporous nanoparticles with different particle size. (e) Effect of $Fe_3O_4@SiO_2$ hollow mesoporous nanoparticles with different particle sizes on the viability of HeLa cells, as measured by MTT assay. (f) Cell viabilities of free DOX, DOX-loaded $Fe_3O_4@SiO_2$, and $Fe_3O_4@SiO_2$ hollow mesoporous nanoparticles at different concentrations. (Reprinted with permission from Zhu Y., Ikoma T., Hanagata N., et al., *Small* 6: 471–478, Copyright 2010, Wiley Publishing Company.)

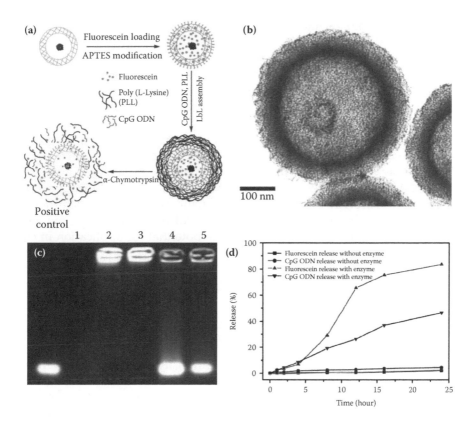

FIGURE 3.8
(a) Schematic procedure for preparation of the fluorescein and CpG ODN-loaded HMS/PLL particles and enzyme-triggered release. (b) TEM images of HMS particles. (c) Gel electrophoresis of the MFHMS suspension and the MFHMS/(CpG/PLL)$_n$ suspensions before and after the treatment with an α-chymotrypsin solution. (d) Release profiles of fluorescein and CpG ODN from the MFHMS/(CpG/PLL)$_3$ particles in 0.1 M acetic acid buffer in the absence and in the presence of α-chymotrypsin (10 μg/mL). (Reprinted with permission from Zhu Y., Meng W., Gao H., et al., *J. Phys. Chem. C.* 115: 13630–13636, Copyright 2011, American Chemical Society.)

gene, and positively charged PLL polymer alternatively to obtain the fluorescein and CpG ODN-loaded HMS/PLL particles (MFHMS/(CpG/PLL)n; n is the assembled number). Finally, fluorescein and CpG ODN can simultaneously release from the MFHMS/(CpG/PLL)n particles stimulated by α-chymotrypsin that can induce PLL polymer degradation. This co-delivery system can give an enzyme-triggered controlled release of drug and gene simultaneously, and the release rates of drug and gene from the HMS/PLL particles can also be controlled by changing the enzyme concentration. Therefore, this system has the advantages of both enzyme-triggered controlled release and co-delivery of drug and gene, and would have potential and promising applications in the field of biomedicine and cancer therapy.

Rattle-type magnetic mesoporous silica nanoparticles have also been prepared through a sol-gel approach associated with water-in-oil microemulsions as templates (Lin et al. 2009; Liu et al. 2010; Tan et al. 2010; Zhang et al. 2011). The nanoparticle morphology, the loading density of magnetic nanocrystals, and the size of the pores can be controlled by adjusting the experimental parameters. Zhang et al. (2008; Liu et al. 2010) reported the preparation of rattle-type periodic mesoporous organosilica magnetic hollow sphere (PMO-MHS) by embedding monodisperse magnetic nanocrystals or large magnetic particles (≈200 nm) into the cavities of highly ordered PMO hollow spheres. In addition, uniform and well-dispersed rattle-type $Fe_3O_4@SiO_2$ nanoparticles of around 50 nm can be prepared through a bio-inspired silicification approach at room temperature and a near-neutral pH aqueous environment (Tan et al. 2010). All these rattle-type magnetic mesoporous silica nanoparticles show the potential for application in controllable drug delivery.

3.3.2 Luminescent Mesoporous Silica Nanoparticles for Controllable Drug Delivery

It is well known that luminescent labeling is a real-time, simple, and effective way to monitor the route of drug-transport carriers in a living system. Drug delivery systems with luminescent labels can easily evaluate the efficiency of the drug release and disease therapy (Insin et al. 2008). As mentioned earlier, mesoporous silica nanoparticles are promising carriers for drug delivery. Therefore, the combination of mesoporous silica and luminescent labels to design luminescent mesoporous silica nanoparticles for drug delivery systems has become a hot research topic.

Traditionally, organic dyes, such as fluorescein isothiocyanate (FITC), rhodamine, and cyanine dyes, are the most common used biological luminescent labels. However, dye molecules usually suffer from photobleaching and quenching when exposed to harsh environments due to interactions with solvent molecules and reactive species such as oxygen or ions dissolved in solution (Wang et al. 2006). Therefore, they are limited for sensitive detection and real-time monitoring. Many efforts have been made to overcome the drawbacks. One of the most promising strategies is to form a core-shell structure that contains a nonporous dye-doped silica core and a mesoporous silica shell, which also processes sustained drug release property and good compatibility (Wang et al. 2009; Lei et al. 2011). As shown in Figure 3.9, Lei et al. (2011) reported a monodisperse multicolor core-shell nanoparticle, using a triple-dye-doped silica nanoparticle as the core and mesoporous silica as the shell. The preparation of multicolor core-shell nanoparticles includes (1) the premodification of the dye molecules: 3-aminopropyltriethoxysilane (APTS) reacts with fluorescein isothiocyanate (FITC), rhodamine B isothiocyanate (RBITC), and rhodamine 101 succinimide (R101-SE) to form three types of dye-APTS conjugates; (2) the hydrolysis and co-condensation of dye-APTS

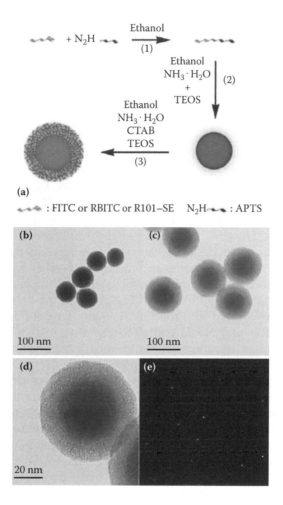

FIGURE 3.9

(a) Synthetic route for multifluorescent core-shell MNSs. TEM images of (b) dye-doped silica cores and (c–d) multifluorescent core-shell MNSs. (e) Confocal fluorescence image of a mixture of eight types of MNSs–IBU under 488 nm argon-ion laser excitation. (Reprinted with permission from Lei J., Wang L., Zhang J., *ACS Nano* 5: 3447–3455, Copyright 2011, American Chemical Society.)

and TEOS to produce dye-doped silica cores; (3) fabricating a mesoporous layer on the dye-doped silica cores to form core-shell nanoparticles by adopting the modified Stöber method. The fluorescent signal of a single multicolor core-shell nanoparticle is about 700 times brighter than its constituent fluorophores. Also, these core-shell nanoparticles showed good drug storage, sustained release capacity, and biocompatibility (Lei et al. 2011). In addition, the researchers developed the co-condensation method with the help of the dehydration reaction between 3-aminopropyltriethoxysilane and dye molecules to prepare dye-doped mesoporous silica nanoparticles, which also possesses high fluorescence quantum yields, excellent photostability, high fluorescence detectivity, sustained drug delivery, noncytotoxicity, and good blood compatibility (He, Shi, et al. 2009; He, Zhang, Chen, et al. 2010). These results are encouraging from the perspective of moving the material platform into clinical trials.

During the past decade, semiconductor nanocrystals, known as quantum dots (QDs), have gained considerable attention as labels in biological imaging and detections due to their outstanding optical properties, such as high quantum yield, excellent photostability, size-dependent tunable fluorescence properties, narrow emission bandwidths, and broad excitation spectra (Peng et al. 1997; Bruchez et al. 1998; Chan and Nie 1998; Mamedova et al. 2001; Michalet et al. 2005; Selvan et al. 2005). Therefore, QDs-based luminescent mesoporous silica nanoparticles have been intensively investigated by many research groups (Hu et al. 2009; Song et al. 2009; Pan et al. 2011). The most common strategy is to design QDs-based luminescent mesoporous silica nanoparticles with QDs as cores and mesoporous silica as shells. Hu et al. (2009) have succeeded in exploiting a facile method to prepare mesoporous silica-coated QDs with multicolor luminescence. In the first step, when QDs in chloroform are added into the CTAB aqueous solution, the CTAB molecules stabilize the oil droplets leading to formation of an oil-in-water microemulsion. Subsequent evaporation of the volatile organic solvent by heating drives CTAB molecules to directly interact with the QD surface ligands through hydrophobic interactions. The alkyl chains of the CTAB and the QD surface ligands intercalate into each other, rendering the CTAB cationic headgroup (quaternary amine) facing outward and the QD-CTAB complex are water soluble. In the second step, when silane compounds polymerize to form silica shells on QD surface, the CTAB acts as templates for mesopore formation. Here, the cationic surfactant molecule, CTAB, plays two important roles in this process: solubilization of hydrophobic QDs into aqueous solutions and templating the mesopore formation (Hu et al. 2009). Therefore, the core-shell luminescent mesoporous silica nanoparticles as drug carriers would open the opportunities in traceable delivery and controlled release of therapeutic agents.

Another excellent luminescent material is rare-earth (RE)-based phosphors, which exhibit intense narrow-band intra-4f luminescence in a wide range of host materials. Furthermore, compared to semiconductor nanocrystals and organic dye molecules, RE-doped nanophosphors have advantages of large Stokes shift, sharp emission spectra, long lifetime, high chemical/photochemical stability, low toxicity, and reduced photobleaching, which can be considered to be a promising luminescent material for biological applications (Nichkova et al. 2005; Goldys et al. 2006). Therefore, luminescent mesoporous silica nanoparticles with RE-doped nanophosphors have been intensively exploited for drug delivery, including embedded structure and a core-shell structure. The embedded structure is to deposit phosphor nanocrystals into the channels of mesoporous silica particles, such as MCM-41,

MCM-48, and SBA-15 (Yan et al. 2007; Yang et al. 2008; Huang et al. 2010). However, this method inevitably blocks the pores to some extent, resulting in the influence on drug loading and release from the carriers.

Recently, more efforts have been made to form a core-shell structure by coating phosphor core with mesoporous silica shell, which reduce the inter-ference of the environment on the luminescent property of the phosphor core and completely utilize the excellent advantages of mesoporous silica matrix. Different RE-doped phosphor cores, such as Y_2O_3:Eu, CeF_3:Tb, Gd_2O_3:Eu, LaF_3:Yb/Er, $NaYF_4$:Yb/Er, and Gd_2O_3:Er, have been coated with mesopo-rous silica shell (Kong et al. 2008; Gai, Yang, Hao, et al. 2010; Yang et al. 2010; Kang et al. 2011; Xu, Gao, et al. 2011; Xu, Li, et al. 2011). All the systems exhibit sustained drug release profiles and excellent luminescent proper-ties. For the core-shell structured up-conversion luminescent and mesopo-rous $NaYF_4$:Yb^{3+}/Er^{3+}@$nSiO_2$@$mSiO_2$ nanoparticles, they were successfully prepared by coating nonporous and further mesoporous silica layers with different thickness on $NaYF_4$:Yb^{3+}/Er^{3+} nanoparticles via a facile two-step sol-gel process as shown in Figure 3.10 (Kang et al. 2011). The nanoparticles exhibit little cytotoxicity, and the *in vitro* release of IBU from the core-shell luminescent mesoporous silica nanoparticles shows a release profile in two steps: an initial diffusion-controlled release, followed by a slower release rate. Furthermore, upon excitation by a 980 nm near-infrared laser, the nanopar-ticles emit green and red fluorescence of Er^{3+} even after the loading of IBU. Interestingly, the drug release amount and process can be monitored by the change of the up-conversion emission intensity. Yang et al. (2010) fabricated LaF_3:Yb^{3+}/Er^{3+}@$nSiO_2$@$mSiO_2$ nanoparticles with a uniform particle diam-eter of 130 nm. The inner luminescence core endues the nanoparticles with up-conversion properties. The outer mesoporous silica shell load IBU in the channels, and most of the incorporated drug molecules can be released to the SBF in 70 h. Therefore, these preliminary results indicate that the luminescent mesoporous silica nanoparticles with a core-shell structure have potential applications in the fields of drug delivery and disease therapy.

3.3.3 Multifunctional Magnetic-Luminescent Mesoporous Silica Nanoparticles for Controllable Drug Delivery

As discussed earlier, both magnetic and luminescent mesoporous silica nanoparticles are of great multifunctional platforms for smart biomedical applications. It can be imagined that the combination of magnetism and luminescence in one mesoporous silica nanoparticle provides a new gen-eration mesoporous silica carrier with a broad range of functionalities. For example, magnetic-luminescent mesoporous silica nanoparticles can serve as an all-in-one diagnostic and therapeutic tool, which could be used to visu-alize and simultaneously treat various diseases.

Yang et al. (2009; Gai, Yang, Li, et al. 2010) designed the core-shell magnetic-luminescent mesoporous silica nanoparticles using nonporous silica-coated

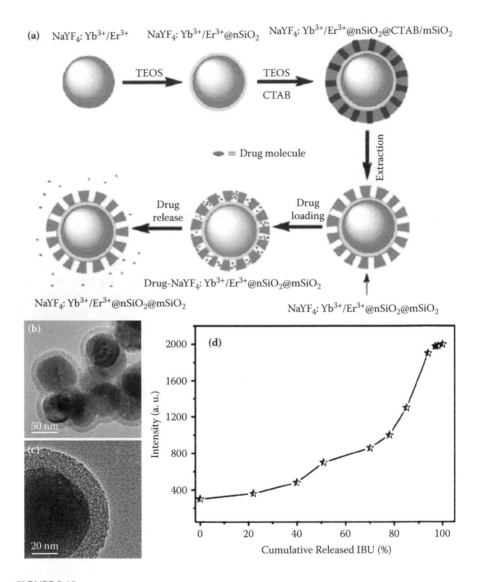

FIGURE 3.10
(a) Schematic for the experimental process for core-shell structured NaYF$_4$:Yb^{3+}/Er^{3+}@nSiO$_2$@mSiO$_2$ nanospheres and subsequent loading and release of IBU. (b–c) TEM images of NaYF$_4$:Yb^{3+}/Er^{3+}@nSiO$_2$@mSiO$_2$ nanospheres. (d) Up-conversion emission intensity of Er^{3+} in IBU- NaYF$_4$:Yb^{3+}/Er^{3+}@nSiO$_2$@mSiO$_2$ as a function of the cumulatively released IBU. (Reprinted with permission from Kang X., Cheng Z., Li C., et al., *J. Phys. Chem. C* 115: 15801–15811, Copyright 2011, American Chemical Society.)

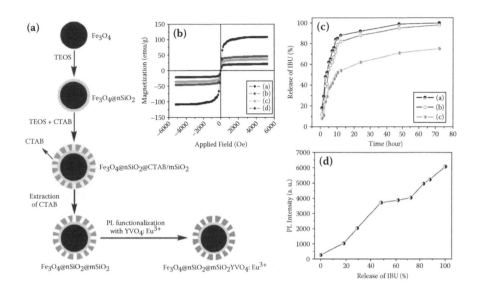

FIGURE 3.11
(a) The formation process of the multifunctional $Fe_3O_4@nSiO_2@mSiO_2@YVO_4:Eu^{3+}$ composite microspheres. (b) Magnetic hysteresis loops of pure (a) Fe_3O_4, (b) $Fe_3O_4@nSiO_2@mSiO_2$, (c) $Fe_3O_4@nSiO_2@mSiO_2@YVO_4:Eu^{3+}$, (d) IBU-$Fe_3O_4@nSiO_2@mSiO_2@YVO_4:Eu^{3+}$. (c) Cumulative IBU releases from the (a) IBU-$Fe_3O_4@nSiO_2@mSiO_2@YVO_4:Eu^{3+}$, (b) IBU-$Fe_3O_4@nSiO_2@mSiO_2$, and (c) IBU-$Fe_3O_4@nSiO_2@mSiO_2@YVO4:Eu^{3+}$-$NH_2$ systems versus release time. (d) PL emission intensity of Eu^{3+} in IBU-$Fe_3O_4@nSiO_2@mSiO_2@YVO_4:Eu^{3+}$ as a function of cumulative released IBU. (Reprinted from Yang P., Quan Z., Hou Z., et al., *Biomaterials* 30: 4786–4795, 2009, with permission from Elsevier Ltd.)

Fe_3O_4 nanospheres as core, ordered mesoporous silica as shell, and further deposition of phosphors (down-conversion $YVO_4:Eu^{3+}$ or up-conversion $NaYF_4:Yb,Er/Tm$ phosphor nanocrystals) on the core-shell nanoparticles (Figure 3.11). These core-shell magnetic-luminescent mesoporous silica nanoparticles exhibit high magnetization, sustained drug release, and excellent luminescent properties. Furthermore, the emission intensity of the multifunctional carrier also increases with the released amount of model drug, thus allowing the release process to be monitored and tracked by the change of photoluminescence intensity. In addition, magnetic-luminescent mesoporous silica nanoparticles were prepared by the combination of magnetism and dye/QDs in mesoporous silica nanoparticles for drug delivery. Liong et al. (2008) and Lee et al. (2010) successfully prepared Fe_3O_4/Dye/$mSiO_2$ core-shell nanoparticles. Here, dye is doped in a mesoporous silica matrix, and Fe_3O_4 nanoparticles are designed as the core in the center or satellites on the surface. Also, Kim et al. (2006) synthesized magnetic-luminescent delivery vehicle using uniform mesoporous silica nanoparticles simultaneously embedded with monodisperse Fe_3O_4 and CdSe/ZnS QDs. Chen, Chen, Zhang et al. (2011) reported the Fe_3O_4/QDs/$mSiO_2$ core-shell nanoparticles for drug delivery with CdTe QDs self-assembled onto the surface of $Fe_3O_4@$

nSiO$_2$@mSiO$_2$ nanoparticles. These results indicated that magnetic-luminescent mesoporous silica nanoparticles can act as a multifunctional drug carrier system, which can target and monitor drugs simultaneously.

3.3.4 Other Multifunctional Mesoporous Silica Nanoparticles for Controllable Drug Delivery

As mentioned before, mesoporous silica can easily combine with other functional materials to form functional mesoporous silica nanoparticles. Except for the magnetic and luminescent mesoporous silica nanoparticles, other functional mesoporous silica nanoparticles also have been intensively developed for drug delivery (Lai et al. 2003; Casasús et al. 2008; Hong et al. 2008; Patel et al. 2008; Du et al. 2009; Gao et al. 2009; Vivero-Escoto et al. 2009; Bernardos et al. 2010; Meng et al. 2010; Sun et al. 2010; Xhen, Zheng, et al. 2010; Chen, Chen, Fang, et al. 2011; Coll et al. 2011; Gan et al. 2011; Xie et al. 2011; Ma et al. 2012). Xie et al. (2011) synthesized the hexagonal mesoporous silica MCM-41 nanospheres with Au nanorods (AuNRs) as core via a modified Stöber method by a process of hydration and condensation of tetraethoxysilane in a water–ethanol mixture. The AuNR@MCM-41 mesoporous silica nanoparticles combine the photothermal characteristic with the mesopore of MCM-41 in one body and could realize the light-driven release of drug due to the photothermal effect of the AuNRs. Therefore, such functional mesoporous silica nanoparticles are favorable for cancer treatment, which combine hyperthermia with the chemotherapeutic drugs by synergistic effect. Ma et al. (2012) reported a novel uniform AuNRs-capped magnetic core/mesoporous silica shell nanoellipsoids (AuNRs-MMSNEs) prepared by coating a uniform layer of AuNRs on the outer surface of a magnetic core/mesoporous silica shell nanostructure, based on a two-step chemical self-assembly process (see Figure 3.12) (Ma et al. 2012). This multifunctional mesoporous silica nanoellipsoids can integrate simultaneous chemotherapy, photothermotherapy, *in vivo* MR, infrared thermal, and optical imaging into one single system. Importantly, the multifunctional nanoellipsoids showed high DOX loading capacity and pH-responsive release, and a synergistic effect of combined chemo- and photothermotherapy could be obtained at moderate power intensity of NIR irradiation based on the DOX release and the photothermal effect of AuNRs.

The aforementioned functional core-shell mesoporous silica nanoparticles are mainly designed with functional cores and mesoporous silica shells. Actually, a variety of functional core-shell mesoporous silica nanoparticles with mesoporous silica cores and functional shells have also been developed for stimuli-responsive controlled drug delivery. That is to say, mesoporous silica cores serve as containers for drugs and functional shells serve as "gate-keepers" to trigger drug release only upon exposure to stimuli, which could decrease side effects to protect the healthy organs from toxic drugs and prevent the decomposition or denaturing of the drugs before reaching the targeted organs or tissues.

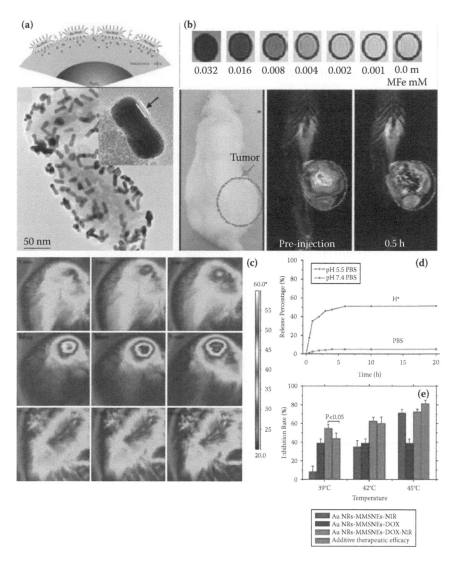

FIGURE 3.12
(See color insert.) (a) Schematic microscopic structure and TEM image of Au NRs-MMSNEs.
(b) T_2 phantom images of Au NRs-MMSNEs at different Fe concentrations and *in vivo* MRI of a
mouse before and after intratumor injection of Au NRs-MMSNEs. (c) Infrared thermal imaging under the photothermal heating by 808-nm laser irradiation for different time periods in
Au NRs-MMSNEs-injected tumor under (up) 1 Wcm^{-2} and (middle) 2 Wcm^{-2} irradiations, and
(down) PBS solution-injected tumor under 2 Wcm^{-2} irradiation. (d) *In vitro* release profiles of
Au NRs-MMSNEs-DOX using dialysis membrane against PBS solution at pH 7.4 and 5.5. (e) A
comparison of inhibition rates for MCF-7 cells treated by Au NRs-MMSNEs-NIR (purple), Au
NRs-MMSNEs-DOX (red), and Au NRs-MMSNEs-DOX-NIR (green). For photothermal treatment, the media were under 808-nm laser irradiation for 5 min at different power intensities,
corresponding to the maximum temperature increases to 39, 42, and 45°C. (Reprinted from Ma
M., Chen H., Chen Y., et al., *Biomaterials* 33: 989–998, 2012, with permission from Elsevier Ltd.)

To date, a variety of stimuli-responsive controlled drug delivery systems based on functional mesoporous silica nanoparticles with mesoporous silica cores and functional shells have been developed (Lai et al. 2003; Casasús et al. 2008; Hong et al. 2008; Patel et al. 2008; Du et al. 2009; Gao et al. 2009; Vivero-Escoto et al. 2009; Bernardos et al. 2010; Meng et al. 2010; Sun et al. 2010; Xhen, Zheng, et al. 2010; Chen, Chen, Fang, et al. 2011; Coll et al. 2011; Gan et al. 2011; Ma et al. 2012). One strategy is to design functional mesoporous silica nanoparticles by coating mesoporous silica nanoparticles with stimuli-responsive polymer layers, biomolecules, or polyelectrolyte multilayers to cap the mesopore channels, such as poly(N-isopropylacrylamide) (PNIPAAm), poly(acrylic acid) (PAA), N-(3-aminopropyl) methacrylamide hydrochloride (APMA), poly(2-(diethylamino)ethyl methacrylate) (PDEAEMA), and poly(methacrylic acid-co-vinyl triethoxylsilane) (PMV) (Hong et al. 2008; Gao et al. 2009; Sun et al. 2010; Xhen, Zheng, et al. 2010). Hong et al. (2008) reported a smart core-shell nanostructure with a mesoporous silica core and a thermoresponsive PNIPAAm nanoshell via surface reversible addition-fragmentation chain transfer polymerization. PNIPAAm nanoshell can be reversibly switched between open and closed states by the change of temperature, which induce the controlled loading and release of drugs from the core-shell mesoporous silica nanoparticles.

Another popular strategy is to design various "gatekeepers" to cap the mesopore outlets of mesoporous silica nanoparticles, such as nanoparticles and supramolecules (Lai et al. 2003; Casasús et al. 2008; Patel et al. 2008; Du et al. 2009; Vivero-Escoto et al. 2009; Bernardos et al. 2010; Meng et al. 2010; Coll et al. 2011; Gan et al. 2011). Using nanoparticles as gatekeepers, solid nanoparticles were chemically attached on the pore outlets of modified mesoporous silica and could be removable with various external stimuli to destroy the chemical bonds, such as pH, redox potential, and temperature (Lai et al. 2003; Vivero-Escoto et al. 2009; Gan et al. 2011). For example, Lai et al. (2003) reported a CdS nanoparticles capped MCM-41 mesoporous silica nanoparticle redox-responsive controlled drug release system. Here, CdS nanocrystals are capped on the mesopore outlets via the disulfide linkages, and the disulfide linkages are chemically labile in nature and can be cleaved with various disulfide-reducing agents, such as dithiothreitol (DTT) and mercaptoethanol (ME). Hence, the release of the CdS nanoparticle caps from the drug-loaded mesoporous silica nanoparticles can be regulated by introducing various amounts of release triggers. A CdS-capped MCM-41 controlled drug release system exhibits less than 1.0% of drug release over a period of 12 h, suggesting a good capping efficiency of the CdS nanoparticles. The addition of DTT disulfide-reducing molecules to the aqueous suspension of CdS-capped mesoporous silica nanospheres triggered a rapid release of the mesopore-entrapped drug, reaching 85% within 24 h, indicating a stimuli-responsive controlled drug release manner for this system.

Macrocyclic organic molecule gatekeepers, a type of supramolecular gatekeeper, can also be disassembled by external stimuli to release the trapped drug molecules. Zink's groups developed several functional mesoporous silica-based pH-responsive controlled drug release systems utilizing the pH-dependent pseudorotaxanes, rotaxanes, or other analogues (Patel et al. 2008; Du et al. 2009; Meng et al. 2010). For example, the aromatic amine stalks were immobilized on the surface of mesoporous silica nanoparticles, and macrocyclic movable gates β-cyclodextrin (β-CD) were introduced to encircle the stalks as a result of noncovalent bonding interactions under neutral pH conditions, and effectively block the nanopore openings for drug storage. Decreasing the pH under mildly acidic conditions leads to protonation of the aromatic amines and dissociation of β-CD caps, following drug models diffusion from the nanopores (see Figure 3.13) (Meng et al. 2010). For liner molecular gatekeepers, stimuli-responsive linear supramolecules are anchored to the external surface of mesoporous silica nanoparticles, and the "close/open" mechanism arises from "across/parallel or shorten" of the liner molecules (Casasús et al. 2008; Bernardos et al. 2010; Coll et al. 2011). For example, Du et al. (2009) tethered the well-known pH-responsive linear polyamine molecules on the pore outlets of MCM-41 nanoparticles through covalent bonds, resulting in both pH-controlled and anion-controlled gate-like effects. The pH-controlled open/close mechanism arises from hydrogen bonding interactions between amines at neutral pH and Columbic repulsions in closely located polyammoniums at acidic pH. The anion-controllable response can be explained in terms of anion complex formation with the tethered polyamines.

3.4 Conclusion and Outlook

In this chapter, the recent research progress on functional mesoporous silica nanoparticles with a core-shell structure for controllable drug delivery was highlighted. A variety of advanced synthetic strategies have been developed to construct functional mesoporous silica nanoparticles with a core-shell structure, including hollow mesoporous silica nanoparticles, magnetic and luminescent core-shell mesoporous silica nanoparticles, and other multifunctional core-shell mesoporous silica nanoparticles. Meanwhile, multifunctional drug delivery systems based on these nanoparticles have been designed and optimized in order to deliver the drugs into the targeted organs or cells, with a controllable release fashion by virtue of various internal and external triggers. The systems will be able to track the released drug molecules in a living system. These developments are encouraging and show great promise in biomedical applications.

However, there are still many challenges that need to be overcome and investigated more comprehensively and thoroughly for these functional mesoporous silica nanoparticles to advance its biological and biomedical

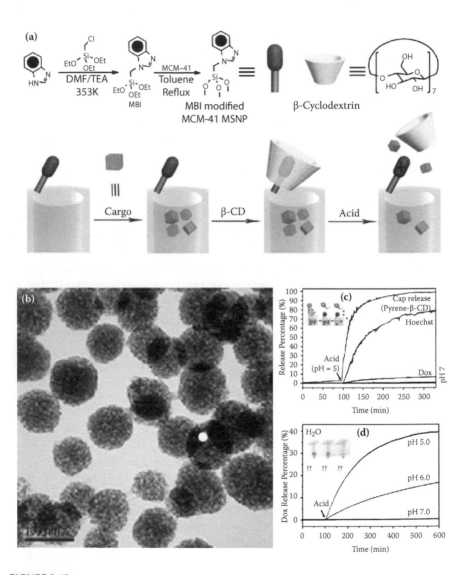

FIGURE 3.13
(See color insert.) (a) A graphical representation of the pH-responsive MSNP nanovalve. (b) TEM image of capped MSNP. (c) Fluorescence intensity plots for the release of Hoechst dye, doxorubicin, and the pyrene-labeled cyclodextrin cap released from MSNP. (d) Release profiles of doxorubicin from ammonium-modified (7.5%, w/w) nanoparticles showing the faster and larger response compared to the unmodified MSNP (c). (Reprinted from Meng H., Xue M., Xia T., et al., *J. Am. Chem. Soc.* 132: 12690–12697, Copyright 2010, American Chemical Society.)

applications. Most of the current mesoporous silica-based drug delivery systems are loaded with small molecular drugs due to the limitation of small mesoporous channels. Actually, large molecular drugs, such as DNA, peptides, and protein, are more important and widely used for therapy. Therefore, it is crucial to design and synthesize multifunctional core-shell mesoporous silica nanoparticles with large mesopore size, which favor the delivery of large molecular drugs. Recently, many studies have reported that mesoporous silica nanoparticles can transport through the cell membranes and deliver drugs into cells, but there are still many questions to be answered for future practical applications, such as the pharmacokinetics and pharmacodynamics of drugs loaded in mesoporous silica nanoparticles, biodistribution, the acute and chronic toxicities, long-term *in vivo* degradation, and compatibility of mesoporous silica nanoparticles. On the other hand, the developed synthesis methods to functional mesoporous silica nanoparticles with a core-shell structure are limited to produce a small amount of nanoparticles. Therefore, the scaled synthesis routes to functional mesoporous silica nanoparticles with a core-shell structure are very important for the final possible biomedical applications.

Acknowledgments

Funding for this study was provided by the Program for Professor of Special Appointment (Eastern Scholar) at Shanghai Institutions of Higher Learning, National Natural Science Foundation of China (No. 51102166), Shanghai Pujiang Program (No. 11PJ1407300), Innovation Program of Shanghai Municipal Education Commission (No. 12ZZ140), Key Project of Chinese Ministry of Education (No. 212055), and the Opening Project of State Key Laboratory of High Performance Ceramics and Superfine Microstructure (No. SKL201004SIC).

References

Andersson J., Rosenholm J., Areva S., et al. 2004. Influences of material characteristics on ibuprofen drug loading and release profiles from ordered micro- and mesoporous silica matrices. *Chem. Mater.* 16: 4160–4167.

Barbé C., Bartlett J., Kong L., et al. 2004. Silica particles: A novel drug-delivery system. *Adv. Mater.* 16: 1959–1966.

Bernardos A., Mondragón L., Aznar E., et al. 2010. Enzyme-responsive intracellular controlled release using nanometric silica mesoporous supports capped with "saccharides." *ACS Nano* 4: 6353–6368.

Bruchez M., Moronne M., Gin P., et al. 1998. Semiconductor nanocrystals as fluorescent biological labels. *Science* 281: 2013–2016.

Casasús R., Climent E., Marcos M. D., et al. 2008. Dual aperture control on pH- and anion-driven supramolecular nanoscopic hybrid gate-like ensembles. *J. Am. Chem. Soc.* 130: 1903–1917.

Chan W. C. W., and Nie S. 1998. Quantum dot bioconjugates for ultrasensitive nonisotopic detection. *Science* 281: 2016–2018.

Chang B., Guo J., Liu C., et al. 2010. Surface functionalization of magnetic mesoporous silica nanoparticles for controlled drug release. *J. Mater. Chem.* 20: 9941–9947.

Chen F., Chen Q., Fang S., et al. 2011. Multifunctional nanocomposites constructed from Fe3O4–Au nanoparticle cores and a porous silica shell in the solution phase. *Dalton Trans.* 40: 10857–10864.

Chen Y., Chen H., Guo L., et al. 2010. Hollow/rattle-type mesoporous nanostructures by a structural difference-based selective etching strategy. *ACS Nano* 4: 529–539.

Chen Y., Chen H., Ma M., et al. 2011. Double mesoporous silica shelled spherical/ ellipsoidal nanostructures: Synthesis and hydrophilic/hydrophobic anticancer drug delivery. *J. Mater. Chem.* 21: 5290–5298.

Chen Y., Chen H., Zeng D., et al. 2010. Core/shell structured hollow mesoporous nanocapsules: A potential platform for simultaneous cell imaging and anticancer drug delivery. *ACS Nano* 4: 6001–6013.

Chen Y., Chen H., Zhang S., et al. 2011. Multifunctional mesoporous nanoellipsoids for biological bimodal imaging and magnetically targeted delivery of anticancer drugs. *Adv. Funct. Mater.* 21: 270–278.

Chen Y., Zheng X., Qian H., et al. 2010. Hollow core-porous shell structure poly(acrylic acid) nanogels with a superhigh capacity of drug loading. *ACS Appl. Mater. Interfaces* 2: 3532–3538.

Coll C., Mondragón L., Martínez-Máñez R., et al. 2011. Enzyme-mediated controlled release systems by anchoring peptide sequences on mesoporous silica supports. *Angew. Chem. Int. Ed.* 50: 2138–2140.

Dekkers S., Krystek P., Peters R. J., et al. 2010. Presence and risks of nanosilica in food products. *Nanotoxicology* 5: 393–405.

Deng Y., Qi D., Deng C., et al. 2008. Superparamagnetic high-magnetization microspheres with an $Fe_3O_4@SiO_2$ core and perpendicularly aligned mesoporous SiO_2 shell for removal of microcystins. *J. Am. Chem. Soc.* 130: 28–29.

Du L., Liao S., Khatib H. A., et al. 2009. Controlled-access hollow mechanized silica nanocontainers. *J. Am. Chem. Soc.* 131: 15136–15142.

Du X., and He J. H. 2011. Facile fabrication of hollow mesoporous silica nanospheres for superhydrophilic and visible/near-IR antireflection coatings. *Chem. Eur. J.* 17: 8165–8174.

Fang X., Chen C., Liu Z., et al. 2011. A cationic surfactant assisted selective etching strategy to hollow mesoporous silica spheres. *Nanoscale* 3: 1632–1639.

Finnie K. S., Waller D. J., Perret F. L., et al. 2009. Biodegradability of sol-gel silica microparticles for drug delivery. *J. Sol-Gel Sci. Technol.* 49: 12–18.

Fu Q. T., He T. T., Yu L. Q., et al. 2010. Preparation and application of silica microspheres with magnetic core/mesoporous silica shell. *Prog. Chem.* 22: 1116–1124.

Gai S., Yang P., Hao J., et al. 2010. Fabrication of luminescent and mesoporous core– shell structured nanocomposites and their application as drug carrier. *Micropor. Mesopor. Mater.* 131: 128–135.

Gai S., Yang P., Li C., et al. 2010. Synthesis of magnetic, up-conversion luminescent, and mesoporous core–shell structured nanocomposites as drug carriers. *Adv. Funct. Mater.* 20: 1166–1172.

Gai S., Yang P., Ma P. A., et al. 2011. Fibrous-structured magnetic and mesoporous Fe3O4/silica microspheres: Synthesis and intracellular doxorubicin delivery. *J. Mater. Chem.* 21: 16420–16426.

Gan Q., Lu X., Yuan Y., et al. 2011. A magnetic, reversible pH-responsive nanogated ensemble based on Fe3O4 nanoparticles-capped mesoporous ilica. *Biomaterials* 32: 1932–1942.

Gao Q., Xu Y., Wu D., et al. 2009. pH-responsive drug release from polymer-coated mesoporous silica spheres. *J. Phys. Chem. C* 113: 12753–12758.

Giri S., Trewyn B. G., Stellmaker M. P., et al. 2005. Stimuli-responsive controlled release delivery system based on mesoporous silica nanorods capped with magnetic nanoparticles. *Angew. Chem. Int. Ed.* 44: 5038–5044.

Goldys E. M., Tomsia K. D., Jinjun S., et al. 2006. Optical characterization of Eu-doped and undoped Gd_2O_3 nanoparticles synthesized by the hydrogen flame pyrolysis method. *J. Am. Chem. Soc.* 128: 14498–14505.

He Q., and Shi J. 2011. Mesoporous silica nanoparticle based nano drug delivery systems: Synthesis, controlled drug release and delivery, pharmacokinetics and biocompatibility. *J. Mater. Chem.* 21: 5845–5855.

He Q., Shi J., Cui X., et al. 2009. Rhodamine B-co-condensed spherical SBA-15 nanoparticles: Facile co-condensation synthesis and excellent fluorescence features. *J. Mater. Chem.* 19: 3395–3403.

He Q., Shi J., Zhu M., et al. 2010. The three-stage *in vitro* degradation behavior of mesoporous silica in simulated body fluid. *Micropor. Mesopor. Mater.* 131: 314–320.

He Q., Zhang J., Chen F., et al. 2010. An anti-ROS/hepatic fibrosis drug delivery system based on salvianolic acid B loaded mesoporous silica nanoparticles. *Biomaterials* 31: 7785–7796.

He Q., Zhang J., Shi J., et al. 2010. The effect of PEGylation of mesoporous silica nanoparticles on nonspecific binding of serum proteins and cellular responses. *Biomaterials* 31: 1085–1092.

He Q., Zhang Z., Gao F., et al. 2011. *In vivo* biodistribution and urinary excretion of mesoporous silica nanoparticles: Effects of particle size and PEGylation. *Small* 7: 271–280.

He Q., Zhang Z., Gao Y., et al. 2009. Intracellular localization and cytotoxicity of spherical mesoporous silica nano- and microparticles. *Small* 5: 2722–2729.

Hirsch L. R., Stafford R. J., Bankson J. A., et al. 2003. Nanoshell-mediated near-infrared thermal therapy of tumors under magnetic resonance guidance. *Proc. Natl. Acad. Sci. USA* 100: 13549–13554.

Hong C.-Y., Li X., and Pan C.-Y. 2008. Smart core-shell nanostructure with a mesoporous core and a stimuli-responsive nanoshell synthesized via surface reversible addition-fragmentation chain transfer polymerization. *J. Phys. Chem. C* 112: 15320–15324.

Hu X., Zrazhevskiy P., and Gao X. 2009. Encapsulation of single quantum dots with mesoporous silica. *Ann. Biomed. Eng.* 37: 1960–1966.

Huang D. M., Hung Y., Ko B. S., et al. 2005. Highly efficient cellular labeling of mesoporous nanoparticles in human mesenchymal stem cells: Implication for stem cell tracking. *FASEB J.* 19: 2014–2016.

Huang S., Li C., Cheng Z., et al. 2012. Magentic Fe3O4@mesoporous silica composites for drug delivery and bioadsorption. *J. Colloid Interface Sci.* 376: 312–321.

Huang S., Li C., Yang P., et al. 2010. Luminescent CaWO4:Tb3+-loaded mesoporous silica composites for the immobilization and release of lysozyme. *Eur. J. Inorg. Chem. 49:* 2655–2662.

Insin N., Tracy J. B., Lee H., et al. 2008. Incorporation of iron oxide nanoparticles and quantum dots into silica microspheres. *ACS Nano* 2: 197–202.

Izquierdo-Barba I., Sousa E., Carlos Doadrio J., et al. 2009. Influence of mesoporous structure type on the controlled delivery of drugs: Release of ibuprofen from MCM-48, SBA-15 and functionalized SBA-15. *J. Sol-Gel Sci. Technol.* 50: 421–429.

Kang X., Cheng Z., Li C., et al. 2011. Core–shell structured up-conversion lumines-cent and mesoporous NaYF$_4$:Yb^{3+}/Er^{3+}@nSiO$_2$@mSiO$_2$ nanospheres as carriers for drug delivery. *J. Phys. Chem. C* 115: 15801–15811.

Kim J., Kim H. S., Lee N., et al. 2008. Multifunctional uniform nanoparticles com-posed of a magnetite nanocrystal core and a mesoporous silica shell for mag-netic resonance and fluorescence imaging and for drug delivery. *Angew. Chem. Int. Ed.* 47: 8438–8441.

Kim J., Lee J. E., Lee J., et al. 2006. Magnetic fluorescent delivery vehicle using uni-form mesoporous silica spheres embedded with monodisperse magnetic and semiconductor nanocrystals. *J. Am. Chem. Soc.* 128: 688–699.

Kong D., Yang P., Wang Z., et al. 2008. Mesoporous silica coated CeF3:Tb3+ particles for drug release. *J. Nanomater.* 1–7.

Kresge C. T., Leonowicz M. E., Roth W. J., et al. 1992. Ordered mesoporous molecular sieves synthesized by a liquid-crystal template mechanism. *Nature* 359: 710–712.

Kumar C. S. S. R., and Mohammad F. 2011. Magnetic nanomaterials for hyperthermia-based therapy and controlled drug delivery. *Adv. Drug Delivery Rev.* 63: 789–808.

Lai C.-Y., Trewyn B. G., Jeftinija D. M., et al. 2003. A mesoporous silica nanosphere-based carrier system with chemically removable CdS nanoparticle caps for stimuli-responsive controlled release of neurotransmitters and drug molecules. *J. Am. Chem. Soc.* 125: 4451–4459.

Laurent S., Dutz S., Häfeli U. O., et al. 2011. Magnetic fluid hyperthermia: Focus on superparamagnetic iron oxide nanoparticles. *Adv. Colloid Interface Sci.* 166: 8–23.

Lee J. E., Lee N., Kim H., et al. 2010. Uniform mesoporous dye-doped silica nanoparti-cles decorated with multiple magnetite nanocrystals for simultaneous enhanced magnetic resonance imaging, fluorescence imaging, and drug delivery. *J. Am. Chem. Soc.* 132: 552–557.

Lei J., Wang L., and Zhang J. 2011. Superbright multifluorescent core-shell mesopo-rous nanospheres as trackable transport carrier for drug. *ACS Nano* 5: 3447–3455.

Li Y., Shi J., Hua Z., et al. 2003. Hollow spheres of mesoporous aluminosilicate with a three-dimensional pore network and extraordinarily high hydrothermal stabil-ity. *Nano Lett.* 3: 609–612.

Li Z., Barnes J. C., Bosoy A., et al. 2012. Mesoporous silica nanoparticles in biomedical applications. *Chem. Soc. Rev.* 41: 2590–2605.

Lian J., Duan X., Ma J., et al. 2009. Hematite (α-Fe$_2$O$_3$) with various morphologies: Ionic liquid-assisted synthesis, formation mechanism, and properties. *ACS Nano* 3: 3749–3761.

Lim J.-S., Lee K., Choi J.-N., et al. 2012. Intracellular protein delivery by hollow meso-porous silica capsules with a large surface hole. *Nanotechnology* 23: 085101.

Lin Y. S., and Haynes C. L. 2009. Synthesis and characterization of biocompatible and size-tunable multifunctional porous silica nanoparticles. *Chem. Mater.* 21: 3979–3986.

Lin Y. S., Wu S. H., Tseng C. T., et al. 2009. Synthesis of hollow silica nanospheres with a microemulsion as the template. *Chem. Commun.* 3542–3544.

Liong M., Lu J., Kovochich M., et al. 2008. Multifunctional inorganic nanoparticles for imaging, targeting, and drug delivery. *ACS Nano* 2: 889–896.

Liu J., Qiao S. Z., Budihartono S., et al. 2010. Monodisperse yolk–shell nanoparticles with a hierarchical porous structure for delivery vehicles and nanoreactors. *Angew. Chem. Int. Ed.* 49: 4981–4985.

Liu Q., Zhang J., Sun W., et al. 2012. Delivering hydrophilic and hydrophobic chemotherapeutics simultaneously by magnetic mesoporous silica nanoparticles to inhibit cancer cells. *Int. J. Nanomed.* 7: 999–1013.

Ma M., Chen H., Chen Y., et al. 2012. Au capped magnetic core/mesoporous silica shell nanoparticles for combined photothermo-/chemo-therapy and multimodal imaging. *Biomaterials* 33: 989–998.

Mamedova N. N., Kotov N. A., Rogach A. L., et al. 2001. Albumin–CdTe nanoparticle bioconjugates: Preparation, structure, and interunit energy transfer with antenna effect. *Nano Lett.* 1: 281–286.

Manzano M., and Vallet-Regí M. 2010. New developments in ordered mesoporous materials for drug delivery. *J. Mater. Chem.* 20: 5593–5604.

Meng H., Xue M., Xia T., et al. 2010. Autonomous *in vitro* anticancer drug release from mesoporous silica nanoparticles by pH-sensitive nanovalves. *J. Am. Chem. Soc.* 132: 12690–12697.

Michalet X., Pinaud F. F., Bentolila L. A., et al. 2005. Quantum dots for live cells, *in vivo* imaging, and diagnostics. *Science* 307: 538–544.

Muñoz B., Rámila A., Pérez-Pariente J., et al. 2003. MCM-41 organic modification as drug delivery rate regulator. *Chem. Mater.* 15: 500–503.

Nichkova M., Dosev D., Gee S. J., et al. 2005. Microarray immunoassay for phenoxybenzoic acid using polymer encapsulated $Eu:Gd_2O_3$ nanoparticles as fluorescent labels. *Anal. Chem.* 77: 6864–6873.

Pan J., Wan D., and Gong J. L. 2011. PEGylated liposome coated QDs/mesoporous silica core-shell nanoparticles for molecular imaging. *Chem. Commun.* 47: 3442–3444.

Park E.-J., Roh J., Kim Y., et al. 2011. A single instillation of amorphous silica nanoparticles induced inflammatory responses and tissue damage until day 28 after exposure. *J. Health Sci.* 57: 60–71.

Patel K., Angelos S., Dichtel W. R., et al. 2008. Enzyme-responsive snap-top covered silica nanocontainers. *J. Am. Chem. Soc.* 130: 2382–2383.

Peng X., Schlamp M. C., Kadavanich A. V., et al. 1997. Epitaxial growth of highly luminescent CdSe/CdS core/shell nanocrystals with photostability and electronic accessibility. *J. Am. Chem. Soc.* 119: 7019–7029.

Rosenholm J. M., Sahlgren C., Linden M., et al. 2011. Multifunctional mesoporous silica nanoparticles for combined therapeutic, diagnostic and targeted action in cancer treatment. *Curr. Drug Targets* 12: 1166–1186.

Rosenholm J. M., Zhang J. X., Sun W., et al. 2011. Large-pore mesoporous silica-coated magnetite core-shell nanocomposites and their relevence for biomedical applications. *Micropor. Mesopor. Mater.* 145: 14–20.

Ruíz-Hernández E., Lopez-Noriega A., Arcos D., et al. 2007. Aerosol-assisted synthesis of magnetic mesoporous silica spheres for drug targeting. *Chem. Mater.* 19: 3455–3463.

Selvan S. T., Tan T. T., and Ying J. Y. 2005. Robust, non-cytotoxic, silica-coated CdSe quantum dots with efficient photoluminescence. *Adv. Mater.* 17: 1620–1625.

Shi H., He X., Yuan Y., et al. 2010. Nanoparticle-based biocompatible and long-life marker for lysosome labeling and tracking. *Anal. Chem.* 82: 2213–2220.

Song Y., Cao X., Guo Y., et al. 2009. Fabrication of mesoporous CdTe/ZnO@SiO$_2$ core/shell nanostructures with tunable dual emission and ultrasensitive fluorescence response to metal ions. *Chem. Mater.* 21: 68–77.

Stöber W., Fink A., and Bohn E. 1968. Controlled growth of monodisperse silica spheres in the micron size range. *J. Coll. Interf. Sci.* 26: 62–69.

Sun J.-T., Hong C.-Y., Pan C.-Y. 2010. Fabrication of PDEAEMA-coated mesoporous silica nanoparticles and pH-responsive controlled release. *J. Phys. Chem. C* 114: 12481–12486.

Tan B., and Rankin S. E. 2005. Dual latex/surfactant templating of hollow spherical silica particles with ordered mesoporous shells. *Langmuir* 21: 8180–8187.

Tan H., Xue J. M., Shuter B., et al. 2010. Synthesis of PEOlated Fe3O4@SiO2 nanoparticles via bioinspired silification for magnetic resonance imaging. *Adv. Funct. Mater.* 20: 722–731.

Tang F. Q., Li L. L., and Chen D. 2012. Mesoporous silica nanoparticles: Synthesis, biocompatibility and drug delivery. *Adv. Mater.* 24: 1504–1534.

Tasciotti E., Plant K., Bhavane R., et al. 2008. Mesoporous silicon particles as a multistage delivery system for imaging and therapeutic applications. *Nat. Nanotechnol.* 3: 151–157.

Uboldi C., Giudetti G., Broggi F., et al. 2012. Amorphous silica nanoparticles do not induce cytotoxicity, cell transformation or genotoxicity in Balb.3T3 mouse fibroblasts. *Mutation Res.* 745: 11–20.

Vallet-Regí M., Balas F., Colilla M., et al. 2008. Bone-regenerative bioceramic implants with drug and protein controlled delivery capability. *Prog. Solid State Chem.* 36: 63–191.

Vallet-Regí M., Rámila A., del Real R. P., et al. 2001. A new property of MCM-41: Drug delivery system. *Chem. Mater.* 13: 308–311.

Van Speybroeck M., Barillaro V., Thi T. D., et al. 2009. Ordered mesoporous silica material SBA-15: A broad-spectrum formulation platform for poorly soluble drugs. *J. Pharm. Sci.* 98: 2648–2658.

Vivero-Escoto J. L., Slowing I. I., Trewyn B. G., et al. 2010. Mesoporous silica nanoparticles for intracellular controlled drug delivery. *Small* 6: 1952–1967.

Vivero-Escoto J. L., Slowing I. I., Wu C.-W., et al. 2009. Photoinduced intracellular controlled release drug delivery in human cells by gold-capped mesoporous silica nanosphere. *J. Am. Chem. Soc.* 131: 3462–3463.

Wang F., Tan W. B., Zhang Y., et al. 2006. Luminescent nanomaterials for biological labelling. *Nanotechnology* 17: R1.

Wang L. Z., Lei J. Y., Zhang J. L. 2009. Building of multifluorescent mesoporous silica nanoparticles. *Chem. Commun.* 16: 2195–2197.

Wu H., Liu G., Zhang S., et al. 2011. Biocompatibility, MR imaging and targeted drug delivery of a rattle-type magnetic mesoporous silica nanosphere system conjugated with PEG and cancer-cell-specific ligands. *J. Mater. Chem.* 21: 3037–3045.

Wu P. G., Zhu J. H., Xu Z. H., et al. 2004. Template-assisted synthesis of mesoporous magnetic nanocomposite particles. *Adv. Funct. Mater.* 14: 345–351.

Xie L., Dong B., Jiang Z., et al. 2011. Synthesis of novel core-shell structural AuNR@ MCM-41 die infrared light-driven release of drug. *J. Mater. Res.* 26: 2414–2419.

Xu Z., Gao Y., Huang S., et al. 2011. A luminescent and mesoporous core-shell structured Gd_2O_3:Eu^{3+}@$nSiO_2$@$mSiO_2$ nanocomposite as a drug carrier. *Dalton Trans.* 40: 4846–4854.

Xu Z., Li C., Ma P. A., et al. 2011. Facile synthesis of an up-conversion luminescent and mesoporous Gd_2O_3:Er^{3+}@$nSiO_2$@$mSiO_2$ nanocomposite as a drug carrier. *Nanoscale* 3: 661–667.

Yan P., Huang S., Kong D., et al. 2007. Luminescence functionalization of SBA-15 by YVO4:Eu^{3+} as a novel drug delivery system. *Inorg. Chem.* 46: 3203–3211.

Yang P., Gai S., Lin J. 2012. Functionalized mesoporous silica materials for controlled drug delivery. *Chem. Soc. Rev.* 41: 3679–3698.

Yang P., Quan Z., Hou Z., et al. 2009. A magnetic, luminescent and mesoporous core-shell structured composite materials as drug carrier. *Biomaterials* 30: 4786–4795.

Yang P., Quan Z., Li C., et al. 2008. Fabrication, characterization of spherical CaWO4:Ln@MCM-41(Ln=Eu3+,Dy3+,Sm3+,Er3+) composites and their applications as drug release systems. *Micropor. Mesopor. Mater.* 116: 524–531.

Yang Y., Qu Y., Zhao J., et al. 2010. Fabrication of and drug delivery by an upconversion emission nanocomposite with monodisperse LaF_3:Yb,ER core/mesoporous silica shell structure. *Eur. J. Inorg. Chem.* 5195–5199.

Yiu H. H. P., Niu H.-J., Biermans E., et al. 2010. Designed multifunctional nanocomposites for biomedical applications. *Adv. Funct. Mater.* 20: 1599–1609.

Zhang C., Hou T., Chen J. F., et al. 2010. Preparation of mesoporous silica microspheres with multi-hollow cores and their application in sustained drug release. *Particuology* 8: 447–452.

Zhang L., Qiao S. Z., Jin Y. G., et al. 2008. Magnetic hollow spheres of periodic mesoporous organosilica and Fe3O4 nanocrystals: Fabrication and structure control. *Adv. Mater.* 20: 805–809.

Zhang X., Clime L., Roberge H., et al. 2011. pH-triggered doxorubicin delivery based on hollow nanoporous silica nanoparticles with free-standing superparamagnetic Fe3O4 cores. *J. Phys. Chem. C* 115: 1436–1443.

Zhao W., Chen H., Li Y., et al. 2008. Cover picture: Uniform rattle-type hollow magnetic mesoporous spheres as drug delivery carriers and their sustained-release property. *Adv. Funct. Mater.* 18: 2780–2788.

Zhao W., Gu J., Zhang L., et al. 2005. Fabrication of uniform magnetic nanocomposite spheres with a magnetic core/mesoporous silica shell structure. *J. Am. Chem. Soc.* 127: 8916–8917.

Zhao W., Lang M., Li Y., et al. 2009. Fabrication of uniform hollow mesoporous silica spheres and ellipsoids of tunable size through a facile hard-templating route. *J. Mater. Chem.* 19: 2778–2783.

Zhou J., Wu W., Caruntu D., et al. 2007. Synthesis of porous magnetic hollow silica nanospheres for nanomedicine application. *J. Phys. Chem. C.* 111: 17473–17477.

Zhu Y., Fang Y., Borchardt L., et al. 2011. PEGylated hollow mesoporous silica nanoparticles as potential drug delivery vehicles. *Micropor. Mesopor. Mater.* 141: 199–206.

Zhu Y., Fang Y., and Kaskel S. 2010. Folate-conjugated Fe_3O_4@SiO_2 hollow mesoporous spheres for targeted anticancer drug delivery. *J. Phys. Chem. C.* 114: 16382–16388.

Zhu Y., Ikoma T., Hanagata N., et al. 2010. Rattle-type Fe_3O_4@SiO_2 hollow mesoporous spheres as carriers for drug delivery. *Small* 6: 471–478.

Zhu Y., Jian D., and Wang S. 2011. Investigation of loading and release of guest molecules from hollow mesoporous silica spheres. *Micro and Nano Lett.* 6: 802–805.

Zhu Y., Kockrick E., Ikoma T., et al. 2009. An efficient route to rattle-type Fe_3O_4@SiO_2 hollow mesoporous spheres using colloidal carbon spheres templates. *Chem. of Mater.* 21: 2547–2553.

Zhu Y., Meng W., Gao H., et al. 2011. Hollow mesoporous silica/poly(L-lysine) particles for co-delivery of drug and gene with enzyme-triggered release property. *J. Phys. Chem. C.* 115: 13630–13636.

Zhu Y., Meng W., and Hanagata N. 2011. Cytosine-phosphodiester-guanine oligodeoxynucleotide (CpG ODN)-capped hollow mesoporous silica particles for enzyme-triggered drug delivery. *Dalton Trans.* 40: 10203–10208.

Zhu Y., and Shi J. 2007. A mesoporous core-shell structure for pH-controlled storage and release of water-soluble drug. *Micropor. Mesopor. Mater.* 103: 243–249.

Zhu Y., Shi J., Chen H., et al. 2005. A facile method to synthesize novel hollow mesoporous silica spheres and advanced storage property. *Micropor. Mesopor. Mater.* 84: 218–222.

Zhu Y., Shi J., Li Y., et al. 2005. Hollow mesoporous spheres with cubic pore network as a potential carrier for the storage of drugs and its in vitro release kinetics. *J. Mater. Res.* 20: 54-61.

Zhu Y., Shi J., Li Y., et al. 2005. Storage and release of ibuprofen drug molecules in hollow mesoporous silica spheres with modified pore surface. *Micropor. Mesopor. Mater.* 85: 75–81.

Zhu Y., Shi J., Shen W., Chen H., et al. 2005. Preparation of novel hollow mesoporous silica spheres and their sustained-release property. *Nanotechnology* 16: 2633–2638.

Zhu Y., Shi J., Shen W., Dong X., et al. 2005. Stimuli-responsive controlled drug release from a hollow mesoporous silica sphere/polyelectrolyte multilayers core-shell structure. *Angew. Chem. Int. Ed.* 44: 5083–5087.

4

3D Plotting of Bioceramic Scaffolds under Physiological Conditions for Bone Tissue Engineering

Yongxiang Luo, Anja Lode, Chengtie Wu, and Michael Gelinsky

CONTENTS

4.1 Introduction

4.1.1 Bioceramic Scaffold-Based Bone Tissue Engineering

Bone loss and osseous defects caused by aging, trauma, tumor resection, or infection severely threaten human health and decrease quality of life. Bone graft, often applied in the clinical situation for treatment of lesions that cannot heal spontaneously, has been recognized as an effective way to repair such defects and regenerate bone function. However, autograft and allograft both have their inherent shortages, such as the limited availability and donor site morbidity for autograft (Paul et al. 2009) or pathogen transmission and immune responses for allograft (Nemzek et al. 1994; Keating and McQueen 2001). Therefore, alternative strategies for bone repair and regeneration involving tissue engineering (TE) approaches are required.

TE aims to generate living and functional tissues or organs *in vitro* by combining cells (such as autologous stem cells isolated from the patient), synthetic scaffolds resembling the natural extracellular matrix, growth-stimulating signals, and bioreactor techniques (Gelinsky et al. 2011; O'Brien 2011; Schroeder and Mosheiff 2011). According to this concept, biomaterials and 3D scaffolds made of them have to provide mechanical support and an environment that facilitates cell attachment, migration, growth, and response to signals (Leong et al. 2008; Srouji et al. 2008; Gelinsky 2009). A broad range of materials including natural and synthetic polymers, bioceramics, and, less important, some biodegradable metal alloys have been developed and widely studied in this field. Compared to metals and polymers, bioceramics have the excellent advantages of osteoconductivity and sometimes even osteoinductivity when used for tissue engineering of bone. Particularly, calcium phosphate (CaP)-based bioceramics and cements (Yuan et al. 1998) consisting of the inorganic component of natural bone have been widely used as bone replacement material as well as for preparation of scaffolds for tissue engineering (Xu et al. 2006; Guo et al. 2009; Dorozhkin 2010).

The porosity of a scaffold is a crucial factor for its success in bone regeneration: complete interconnectivity of the pores as well as pore sizes in an appropriate range are necessary for cell migration, new bone ingrowth, and vascularization (Karageorgiou and Kaplan 2005; Jones et al. 2009). On the other hand, the porous structure should still be able to mediate mechanical stability adequate for the host tissue. Several methods have been developed and are used for the preparation of porous scaffolds. The most commonly used technology for fabrication of CaP and other bioceramic scaffolds is particle leaching. In this method, either water-soluble materials (such as NaCl) or those dissolvable in hydrophobic organic solvents (like paraffin) are incorporated as porogens into CaP pastes. After shaping, the CaP scaffolds are incubated in water for setting and dissolving the porogens that lead to mechanical loadable porous structures, or are treated with an organic solvent after setting to remove the hydrophobic porogens (Guo et al. 2009). An alternative way is the utilization of a porous polymer template (such as polyurethane foams), which is coated with a CaP slurry. By sintering that structure at high temperatures, a stable, porous CaP scaffold is achieved while the polymer component is burned out (e.g., Zhang and Zhang 2002). However, these methods are limited by poor control of pore size and morphology as well as insufficient pore interconnectivity. Therefore, in the past decade, advanced techniques of rapid prototyping (RP) based on computer-aided design (CAD) and computer-aided manufacturing (CAM) were introduced in the field of tissue engineering and regenerative therapies (Landers et al. 2002; Yang et al. 2002).

4.1.2 Rapid Prototyping Techniques Applied for Scaffolds Fabrication

The principle of rapid prototyping, also known as solid free form fabrication, is the generation of scaffolds through layer-by-layer construction of 3D structures with predefined inner and outer geometry. Basis for the CAD/CAM data sets can be computer tomography or magnetic resonance tomography of the defect region, which are used to generate a virtual 3D model that is then converted into a sequence of slices that are used to build the corresponding real 3D object in layer-by-layer fashion (Hutmacher et al. 2004; Pfister et al. 2004). In contrast to conventional methods used for scaffold fabrication, the preparation of molds as well as subsequent machining steps for shaping are not necessary. Thus, the main advantage of RP techniques is the possibility to design and control the architecture of a scaffold, not only with respect to the external geometry predetermined by size and shape of the patient-specific defect but also a defined pore structure including channels for nutrient supply and vascularization can be realized.

A number of RP techniques has been successfully adapted for processing of biomaterials into 3D scaffolds such as systems based on laser technology (stereolithography and selective laser sintering), devices using printing technology (3D powder printing), and extrusion-based systems (Hutmacher et al. 2004; Pfister et al. 2004). 3D plotting as well as fused deposition modeling, which belong to the last group, function by dispensing strands of a pasty or molten material through a moving nozzle to build a 3D structure in layered fashion (Hutmacher et al. 2004). However, whereas fused deposition modeling is restricted to thermoplastic materials with good melt viscosity, a wide variety of synthetic and natural materials can be processed by using the technique of 3D plotting.

Examples for materials that can be used for scaffold preparation applying the 3D plotting technique, originally developed by Muehlhaupt and coworkers at the Freiburg Materials Research Center (Germany), are hydrogels (Landers et al. 2002; Fedorovich et al. 2008; Maher et al. 2009), hydroxyapatite (HA) ceramic slurries (Detsch et al. 2008), polycaprolactone (Kim and Son 2009; Oliveira et al. 2009), and starch-based blends (Martins et al. 2009). In contrast to other 3D extrusion technologies, the mild process conditions (room or physiological temperature and no usage of organic solvents) allow simultaneous plotting of biological components such as antibiotics, growth factors, and even living cells that are suspended in the respective plotting material (Fedorovich et al. 2008; Lode et al., forthcoming).

4.1.3 Rapid Prototyping of Bioceramics

A lot of work has been invested in the fabrication of CaP scaffolds by 3D printing, also referred to as powder printing. In this process, calcium phosphate particles are bonded using a liquid binder that is delivered by an inkjet printing head (Sachs et al. 1993). After the printing process, the binder-free powder is removed and the remaining structure is stabilized, in most cases by sintering, which results in a ceramic body possessing high crystallinity and mechanical stability (Seitz et al. 2005; Khalyfa et al. 2007). Using this technique, scaffolds with designed pore parameters have been successfully fabricated. However, there are still some disadvantages such as difficulty removing the unbound powder, especially from small pore structures.

As an alternative strategy, 3D extrusion approaches of CaP pastes and slurries have been developed. Miranda and coworkers have used so-called ceramic inks, highly concentrated, water-based suspensions of β-tricalcium phosphate or HA powder, which are dispensed through a moving deposition nozzle in an oil bath to build a ceramic scaffold (Miranda et al. 2006, 2007, 2008; Franco et al. 2010). This process, introduced in the literature as "robocasting" or "direct write assembly," is very similar to the technique of 3D plotting. After building, the scaffolds are dried and sintered at high temperature resulting in ceramic bodies. However, the second sintering hinders loading of drugs, growth factors, and living cells in the CaP pastes.

4.1.4 Rapid Prototyping-Based Tissue Engineering

There are three strategies to engineer tissues, tissue substitutes, or even organs by using RP technologies (Figure 4.1). According to the first, the scaffold is fabricated by processing an adequate biomaterial into 3D porous scaffolds with a predesigned outer and inner structure. In principle, each RP method described earlier can be used if it is suitable for the respective material. After fabrication, the scaffold can be modified to improve its properties. Subsequently, living cells (e.g., stem cells) are seeded and cultured on the sterilized scaffolds *in vitro* and the construct finally is implanted *in vivo*. The main advantage of this strategy is the nearly unlimited feasibility to modify the scaffolds, because living cells are added only after scaffold fabrication is completed. For example, biomaterials and scaffolds can be modified by introducing active groups or improving mechanical properties under chemical reaction and heat treatment during processing and postprocessing. However, the disadvantage of this strategy is that cell seeding is relatively uncontrolled and often associated with a low seeding efficiency and inhomogeneous cell distribution. Although a number of upgraded cell seeding

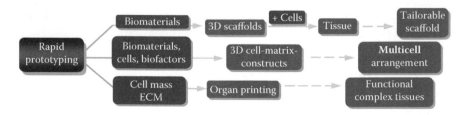

FIGURE 4.1
Strategies of rapid prototyping-based tissue engineering.

techniques has been developed resulting in an improvement of cell seeding efficiency and distribution, such as semidynamic or dynamic cell seeding techniques, using rotating or shaking systems, spinner flasks, or bioreactors, this strategy has general limitations with respect to seeding of more than one cell type or seeding with defined location or distribution of the cells.

One of the most exciting features of the RP technique 3D plotting is the possibility to integrate delicate biological compounds and even cells into a scaffold during the fabrication process. Therefore, this technique is suitable to realize the second strategy, shown in Figure 4.1, which comprises the simultaneous processing of living cells and biomaterials into cell–matrix constructs. This is realized by mixing cells, optionally growth factors, or other supporting substances, and the pasty biomaterial prior to scaffold fabrication followed by plotting of these multicomponent pastes under sterile conditions. A prerequisite for this strategy is the possibility to stabilize the biomaterial after plotting in a cell-compatible way, which is the most limiting factor. In addition, the plotting material meeting the stringent requirements with respect to cell compatibility also has to match the mechanical stiffness of the respective extracellular matrix (ECM). The plotted cell–matrix construct is incubated in cell culture medium *in vitro* and finally transplanted *in vivo*. The main advantages of this strategy comprise not only the high cell seeding efficiency and homogeneous distribution of the cells, but also the ability to generate complex composed scaffolds by incorporation of several cell types in a spatially controlled manner, optionally in combination with more than one biomaterial. This can be realized by using devices equipped with a multichannel dosing system (Lode et al., forthcoming). For example, the fabrication of biphasic scaffolds suitable for regeneration of lesions at tissue interfaces such as osteochondral defects and the creation of prevascularized tissue equivalents (Luo, Lode, et al. forthcoming) are rewarding tasks that can be achieved with this strategy.

The final and most ambitious strategy is called "organ printing." Organ printing is a version of using RP to fabricate (pieces of) organs that can be transplanted into patients, especially for those complex organs such as the heart, liver, and kidney. This technique, whose concept is—in contrast to the aforementioned strategies—the creation of functional tissues and organs without solid biodegradable scaffolds, can be defined as layer-by-layer biofabrication of macrotissues by using tissue spheroids as building blocks. After connecting to each other, the tissue spheroids can fuse, forming macrotissues and even organs in a self-assembly process (Mironov et al. 2009). The RP technique capable of realizing the precise 3D positioning of several cell types is inkjet printing with tissue spheroids as processable "bioinks." Computer-aided robotic devices ("bioprinters") with multiple nozzles dispense the bioinks precisely in a defined order onto sprayed hydrogel layers ("biopaper")—a process that is repeated layer by layer to build a 3D tissue construct (Mironov et al. 2009). However, organ printing remains a big dream and various problems have to be solved in the future.

In this chapter, 3D plotting of diverse bioceramic scaffolds mainly based on calcium phosphate phases for bone tissue engineering will be introduced. The mechanical properties and cell culture *in vitro* have been evaluated. Furthermore, a polymer reinforced calcium phosphate scaffold will also be discussed as well as its improved mechanical properties and cell behavior. In addition, 3D plotting of silicon-based bioceramic scaffolds are presented, including the significantly improved mechanical properties and toughness.

4.2 3D Plotting of Calcium Phosphate Cement (CPC) Scaffolds

As introduced earlier, calcium-phosphate-based materials are the distinguished candidates for bone repair and regeneration because of their excellent properties, such as osteoconductivity, and due to the fact that calcium phosphates are the most important inorganic constituents of biological hard tissues. This part describes the fabrication of calcium phosphate scaffolds by 3D plotting using an optimized calcium phosphate cement (CPC) paste. Plotting and postprocessing steps are conducted without any heat treatment, change of pH value, or involvement of organic solvents that allows—in contrast to other RP techniques used for production of calcium phosphate scaffolds such as powder printing or selective laser sintering—the incorporation of pharmaceutical or biological components like antibiotics, drugs, growth factors, or even autologous constituents (such as blood components or living cells) during scaffold fabrication. Hence, this approach seems to be suitable to realize individualized therapies that meet the demands of the patients not only concerning implant geometry but also with regard to the biological aspects.

Figure 4.2 illustrates the process of CPC scaffold fabrication by 3D plotting including the postprocessing steps. A ready-to-use CPC paste (P-CPC) is filled in a cartridge that is loaded in the plotting unit. By using compressed air, P-CPC strands are extruded and laid down to build a CPC scaffold with predesigned structure (Figure 4.2a,b). After plotting, the CPC scaffolds are transferred into water or buffered aqueous solutions for setting and hardening (Figure 4.2c). After drying at room temperature, the CPC scaffolds can be sterilized by gamma irradiation prior to their application (Figure 4.2d,e). Alternatively, the process can be completely performed under sterile conditions, with a setting in a cell culture medium at 37°C and without drying in the case of incorporation of living cells or with hardening not in an aqueous solution but in a humidified atmosphere in case of loading with growth factors or drugs. Printed CPC scaffolds with different pore sizes and morphologies can easily be achieved (Figure 4.2f).

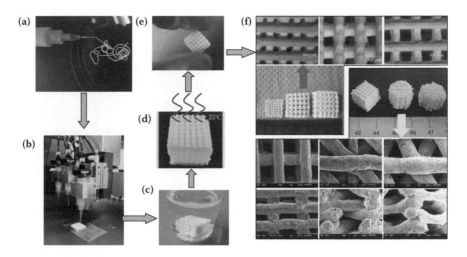

FIGURE 4.2
The process of CPC scaffold fabrication by 3D plotting. The P-CPC paste used for plotting was filled in a cartridge and tested for its printability: (a) uniform CPC strands can be extruded through a stainless steel needle. (b) A 3D porous scaffold with designed structure is fabricated by 3D plotting; (c) after completion of plotting, the scaffold is transferred into water for setting and hardening. (d) Afterward, the scaffold is dried at room temperature and (e) is ready for further application. (f) Scaffolds with different pore size and morphologies can be produced.

4.2.1 Preparation of the Plotable CPC Paste and 3D Plotting of CPC Scaffolds

A prerequisite for plotting of CPC scaffolds under physiological conditions is the development of suitable CPC pastes. Conventional powder/liquid CPC pastes are prepared by mixing solid calcium phosphate cement precursors with water or aqueous Na_2HPO_4 solution resulting in an injectable paste that sets and hardens within a few minutes. This type of CPC paste is suitable and frequently utilized for filling of bone defects in clinical applications; however, the fast setting reaction and hardening prevents its usage for scaffold fabrication by 3D plotting since the powder/liquid CPC would block the nozzle with time. This limitation is circumvented by the application of a newly developed CPC paste (P-CPC; patent applied by InnoTERE, Dresden, Germany) that can be stored as a malleable paste for months. The P-CPC is prepared by mixing 80% of the cement precursor powder with 20% of a carrier liquid consisting of a semisynthetic biocompatible oil (short chain triglycerides with 8–12 C atoms) and emulsifiers (Lode et al., forthcoming). The cement powder consists of 60% α-tricalcium phosphate (α-TCP), 26% dicalcium phosphate anhydride (DCPA), 10% $CaCO_3$, and 4% precipitated HA (Khairoun et al. 1997). The setting reaction of the P-CPC is triggered by contact with water or aqueous solutions such as buffer and cell culture medium *in vitro* or body liquids *in vivo*. Due to these favorable characteristics, the injectability of the P-CPC is fully maintained during the plotting

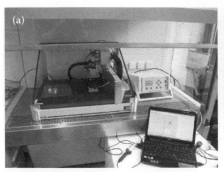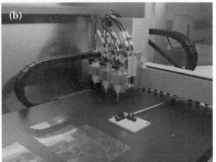

FIGURE 4.3
(a) The 3D plotting system including the 3D plotting device, compressed air controller, and computer to govern the CAD/CAM process. The plotting device is located in a laminar flow box for operation under sterile conditions. (b) The plotting unit comprises three cartridges enabling fabrication of scaffolds consisting of three different pasty materials or additional components like drugs, growth factors, or living cells.

process. After completion of 3D scaffold fabrication, the plotted structure is transferred into water or other aqueous solutions (e.g., phosphate buffered saline, cell culture medium) to start the setting reaction. The CPC scaffolds are hardened within 10 min, but the scaffolds are kept in aqueous media for a couple of days. During this time, the cement precursors are transformed to hydroxyapatite, and the oil, which is used for preparation of the storable P-CPC, is slowly displaced.

3D plotting of CPC scaffolds was carried out in a CAD/CAM process applying the plotting system shown in Figure 4.3. The Bio-Scaffold-Plotter, developed at the Fraunhofer IWS Dresden (Germany) on the basis of the Nano-Plotter™ from GeSiM (Groerkmannsdorf, Germany), consists of three main parts: a plotting device, an air-pressure controller, and a computer with software required to control the CAD/CAM process (Figure 4.3a). The plotting device has a platform and a plotting unit with three channels (Figure 4.3b) that is guided by a high-precision three-axis motion system allowing movement in the x-, y-, and z-directions. The three channels of the plotting unit enable simultaneous processing of up to three different pasty materials or mixtures within one scaffold or implant. Beside the generation of patient-specific implants by incorporation of biological components, bi- or triphasic scaffolds for repair and regeneration of defects at tissue interfaces can be produced (Lode et al., forthcoming).

For plotting, a cartridge equipped with a fine nozzle is filled with P-CPC, inserted into the plotting unit, and connected with a compressed air supply. During extrusion of the pasty material forced by compressed air, the nozzle accomplishes a 3D motion according to the CAD/CAM data set. The plotting unit moves with a constant speed (plotting speed) along the x- or y-axis or intermediate x- or y-directions. If plotting of one strand is completed, extrusion of the material is stopped and a high tear off speed is applied to disrupt the strand at the end. The plotting unit is lifted and moved to the start point of the next strand. If all strands of one layer are deposited, the plotting unit is moved along the z-axis to start plotting of the next layer. This change of the position in z-direction is also controlled by the CAD/CAM data set and depends on different parameters such as the inner diameter of the extrusion nozzle, material properties, air pressure, and plotting speed, which in turn affect the strand dimensions. The process is repeated until the whole scaffold is generated. Size and shape of the resulting structure is predefined by the CAD data set but depend also on the dosing air pressure, plotting speed, and the diameter of the dosing nozzle.

Working air pressure and plotting speed have to be carefully adapted to each other and optimized for the used nozzle diameter to ensure the extrusion and deposition of homogeneous strands. Figure 4.4 demonstrates their influence on strand width: an increase of the plotting speed and a decrease of the air pressure result in a decreased width of plotted strands, an observation that has been made for different needle diameters. If the speed is too

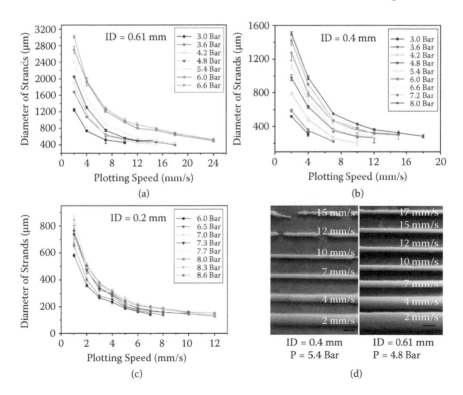

FIGURE 4.4
The relationship between plotting speed and diameter of plotted strands as a function of air pressure. Different nozzles were used with inner diameters (ID) of (a) 0.61 mm, (b) 0.4 mm, and (c) 0.2 mm. (d) Photographs of CPC strands plotted under different air pressures and plotting speeds.

low and the pressure too high, strands are generated that are much wider than the nozzle diameter; the reverse conditions (high speed and low pressure) result in the production of inhomogeneous strands that can be disrupted (Figure 4.4d).

Preparation parameters of the P-CPC such as powder-to-oil ratio and particle size have a strong influence on injectability of the P-CPC as well as on the properties of the plotted scaffolds. For instance, a P-CPC with a lower powder-to-oil mass ratio can be extruded through a finer nozzle with an inner diameter of 150–250 μm, however, the material is not stiff enough to support the whole weight of the plotted scaffold. This results in closure of the pores that are oriented parallel to the building platform and deformation of the scaffold, especially for those with a height in the centimeter range that might be clinically relevant. To solve this problem, two strategies can be considered. One is to increase the powder-to-oil ratio of the pastes, which can enhance the stiffness of the P-CPC, but with the compromise to accept a decrease in the extrusion ability and continuity by possible blocking of the plotting nozzles, especially for fine nozzles (100–250 μm in diameter), which are necessary for achieving scaffolds with high resolution.

The other way is the application of a temporary material for filling the space between the CPC strands and therefore supporting the structure during the plotting process. This sacrificial material should be injectable to enable plotting with the same device and water soluble so that it can be dissolved during setting and hardening of the CPC scaffold in water to unblock its macroporous structure. We successfully adopted a 25 wt% polyvinyl alcohol (PVA) paste as temporary material since it fulfills these requirements and obtained CPC scaffolds that kept their shape and whose pores oriented parallel to the x- and y-axis were completely opened and uniform. In contrast, the pores of scaffolds fabricated from the same P-CPC material but without temporary PVA strands were partially closed at the cross-section direction (Figure 4.5). The stabilizing effect of the PVA supporting strands is not only caused by simple filling of the space between two CPC strands, but also due to the fact that the PVA paste contains water that permeates the CPC strands and triggers the cement setting reaction. This phenomenon results in a graded increase of stiffness of the layers from the bottom up as a function of contact time between CPC strands and water from PVA sol.

The application of a sacrificial paste for plotting of supporting strands, which was successfully used for 3D plotting of CPC scaffolds with completely interconnected pores, might also be interesting for 3D plotting of other scaffolds such as silicon-based bioceramics, biopolymer hydrogels, or composites.

Deformation of the scaffolds including closure of the macropores in x-/y-direction described earlier (Figure 4.5b) is caused by gravitational force, which is effective especially in case of a considerable difference in the density of plotting material (P-CPC) and plotting medium (air). Therefore, an alternative method might be plotting of P-CPC into water or cell culture medium to minimize this difference in density. In this case, the CPC strands

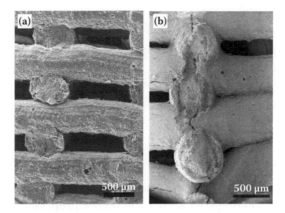

FIGURE 4.5
Scanning electron microscopy (SEM) images of plotted CPC scaffolds (a) with and (b) without application of PVA paste as supporting material (side view).

would harden immediately after extrusion and thereby achieve an increased stiffness that should be sufficient to support the weight of the whole scaffold resulting in uniform and open pores in the x-/y-direction. However, problems concerning the weak bonding between several layers and potential blocking of the nozzle are still unresolved.

4.2.2 Characterization of the Plotted CPC Scaffolds

CAD/CAM enables the fabrication of CPC scaffolds with different size and shape by using the 3D plotting technique, not only with respect to the overall shape that fits into a patient-specific defect but also with an optimized inner pore size and structure to support cell attachment and migration as well as blood capillary ingrowth on the one hand and to mediate adequate mechanical stability on the other hand. As shown in Figure 4.2f, CPC scaffolds with designed pore size and morphologies can be fabricated by using the 3D plotting technology illustrating the key advantage of RP techniques—compared to conventional methods used for scaffold preparation—namely, the control and optimal adjustment of pore parameters such as size, interconnectivity, distribution, and morphology. The strong influence of size and morphology of the pores on scaffold performance concerning mechanical properties, cellular response, and new bone formation *in vivo* has been demonstrated in many publications. Some studies have revealed that scaffolds with pore size around 300 to 400 µm were optimal for bone tissue engineering (Tsuruga et al. 1997; Murphy et al. 2010). In addition, a complete interconnectivity of the pores is also necessary for cell growth and migration, and homogeneous new bone formation in the scaffold.

One important issue with respect to tissue engineering applications is the resolution of plotted scaffolds, as it has a strong influence on pore size and structure. The resolution depends mainly on characteristics of the basal

FIGURE 4.6
Light microscope images of CPC scaffolds plotted with nozzle diameters of (a) 200 µm and (b) 410 µm. (c) A biphasic scaffold consisting of two parts with different porosities.

material (e.g., molecular weight of polymers, particle size of ceramic powder, composition of composites, etc.) and on the properties of the respective pastes prepared for plotting (such as viscosity and homogeneity). Critical factors and parameters that affect the viscosity and therefore the resolution of plotted scaffolds are the solid-to-liquid ratio and temperature (in case of polymer melts). In our work, the cement particle size has been reduced by optimization of the grinding procedure during preparation of the P-CPC that resulted in a clear enhancement of the resolution achieved by application of finer nozzles: CPC strands of less than 200 µm width can be plotted.

The microscopic images depicted in Figure 4.6 demonstrate the performance of the new technique of CPC plotting for scaffold fabrication. Plotting of the optimized P-CPC with nozzle diameters of 200 and 410 µm, respectively, lead to a uniform pore size and consistent pore structure. Moreover, CPC scaffolds possessing an anisotropic structure—one example is shown in Figure 4.6c—can also be fabricated with this technique. In that scaffold, a denser part with smaller pores and lower porosity for mimicking compact bone and a less dense part with bigger pores and higher porosity mimicking trabecular bone were designed and fabricated, leading to a distinct anisotropic material. In addition, some studies have shown that scaffolds with small pores can lead to hypoxic conditions and induce osteochondral tissue formation before osteogenesis, while large pores, that can be better vascularized, lead to direct osteogenesis (Karageorgiou and Kaplan 2005). Therefore, such anisotropic structures with gradients in pore size would be recommended for the formation of multiple tissues and tissue interfaces such as biphasic scaffolds for repair of osteochondral defects.

Scaffolds developed for tissue engineering as well as implants should be able to mediate stability to the defect ideally possessing mechanical properties comparable to those of the host tissue. Plotted CPC scaffolds of different porosity (generated by using different strand distances) were tested for compressive strength and modulus. The data revealed that plotted CPC scaffolds with completely interconnected pores have compressive strengths of 1 to 3 MPa, reciprocally proportional to the porosity, that is, the mechanical stability decreases with increase of porosity (Figure 4.7). For porous scaffolds, the mechanical properties do not only depend on the strength of the bulk

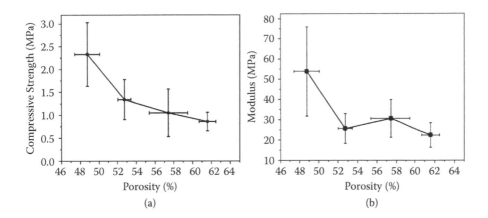

FIGURE 4.7
(a) Compressive strength and (b) modulus of plotted CPC scaffolds with different porosity (n = 5). Scaffolds with a size of 10 mm × 10 mm × 10 mm were tested by using an Instron 5566 testing machine with a 10 kN load cell and a constant rate of 1 mm/min.

material but are also strongly affected by the pore parameters including size, morphology, distribution, and orientation. In case of scaffolds fabricated by RP technologies, the influence of macroporosity and scaffold architecture, predefined by the CAD data set, on mechanical strength has been clearly demonstrated by several studies (Hutmacher et al. 2001; Zein et al. 2002; Sobral et al. 2011). The mechanical properties of the plotted CPC scaffolds are, to a certain extent, adaptable by design and changing the pore parameters. Generally, a uniform and continuous pore structure and orientation improves the mechanical strength (Wu et al. 2008, 2011) and in this respect, scaffold fabrication applying the technique of 3D plotting has an advantage over other conventional methods.

The 3D plotted CPC scaffolds set in water were mechanically weaker than those sintered at high temperature. We studied the modulus of the plotted CPC scaffolds with and without sintering and found that the plotted CPC scaffolds produced with an additional sintering step (1250°C for 3 h) acquired a modulus of about 330 MPa, which is nearly seven times higher than that of samples with the same geometry produced without sintering (Figure 4.8a). As apparent from SEM investigations (Figure 4.8b,c), high temperature sintering results in highly crystalline materials with, compared to cements, bigger particles that are connected to each other by sinter necks. This structural feature is accompanied by a higher mechanical strength. In contrast, CPC consists of smaller, mostly nanocrystalline particles that were formed and became entangled during the precipitation reaction providing mechanical rigidity (Bohner 2000). Accordingly, the SEM image shown in Figure 4.8a reveals that the surface of the nonsintered CPC strands is highly porous with plenty of small particles that seem to be bonded weaker compared to the more dense counterparts produced by sintering.

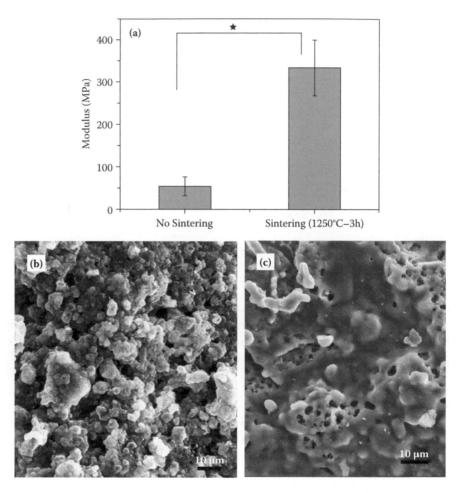

FIGURE 4.8
High-magnification SEM images showing the microstructures of (b) nonsintered and (c) sintered (at 1250°C and 3 h) CPC scaffolds and (a) the Young's modulus of the two types of scaffolds.

Most porous calcium phosphate bioceramic scaffolds are produced or at least posttreated at high temperatures since the high crystallinity and increased density induced by the sintering strongly enhance the mechanical properties. On the other hand, the biodegradability of such materials is limited (Bohner 2000) and the incorporation of drugs, growth factors, or autologous components is possible only after fabrication of the scaffold. CPC does not require postsintering to achieve stable structures and it can set and harden under mild conditions, according to the requirements in water or a humidified environment. In addition, due to free spaces between precipitated crystals, calcium phosphate bone cements exhibit a high intrinsic microporosity and the resultant large surface area is beneficial for drug loading (Espanol et al. 2009). The nanocrystalline structure of CPC finally allows their resorption by osteoclasts and remodeling into new bone tissue (Constantz et al. 1995; Jansen et al. 1995).

4.2.3 *In Vitro* Biological Evaluation of the CPC Scaffolds

An important feature of a scaffold developed for tissue engineering applications is its performance as a cell carrier. We have seeded human mesenchymal stem cells (hMSC), derived from bone marrow, on the plotted CPC scaffolds to evaluate their cytocompatibility (Lode et al., forthcoming). The cell-seeded scaffolds were cultivated over 19 days in cell culture medium containing osteogenic supplements (dexamethasone, β-glycerophosphate, and ascorbic acid 2-phosphate), which are able to induce the differentiation of hMSC into osteoblasts *in vitro*. SEM analyses revealed that the cells attached to and spread over the surface of the plotted CPC scaffolds (Figure 4.9a,b).

After longer cultivation time, the cells proliferated and migrated, even bridging the gaps between the CPC strands (Figure 4.9c) and finally formed a thick cell sheet (Figure 4.9d). Proliferation and differentiation along the osteoblastic lineage were evaluated by biochemical determination of the enzyme activities of lactate dehydrogenase (LDH), a housekeeping enzyme whose activity can be correlated with the number of living cells, and of alkaline phosphatase (ALP), a typical marker of the early stage of osteoblastic differentiation. Cytosolic LDH activity measurement revealed that the number of viable cells increased over the cultivation period (Figure 4.9e). The rise of the specific ALP activity with a maximum value at day 12 strongly indicates the osteogenic differentiation of hMSC cultivated on the CPC scaffolds in the presence of osteogenic supplements (Figure 4.9f). The data of the cell experiment demonstrated the capability of the plotted CPC scaffolds to support hMSC attachment, growth, and osteoblastic differentiation.

4.3 Polymer Reinforced Biphasic CPC–Alginate Scaffolds

Ceramic and polymer scaffolds developed for bone tissue engineering applications have, defined by their intrinsic properties, specific advantages and disadvantages. For instance, the low mechanical strength and inherent brittleness of pure calcium phosphate scaffolds prepared without sintering remain limitations of their clinical application. The development of ceramic and polymer composite scaffolds is a promising option to generate structures of enhanced quality with respect to their mechanical properties (Gelinsky and Heinemann 2010). Such organic and inorganic composite scaffolds mimic the natural bone to some extent since bone is likewise composed of an inorganic (hydroxyapatite) and an organic biopolymer part (mostly

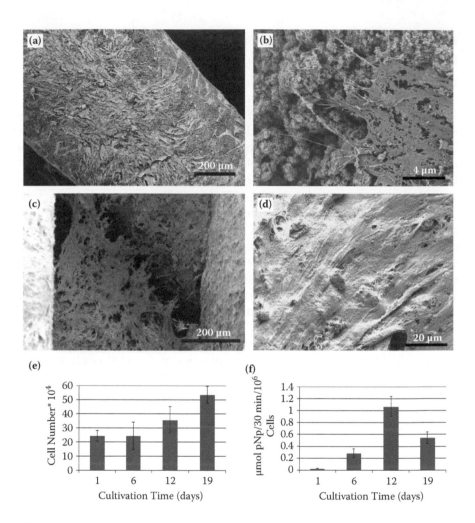

FIGURE 4.9
Cultivation of hMSC on plotted CPC scaffolds. Prior to cell seeding, the scaffolds were sterilized by gamma radiation and preincubated in cell culture medium for 24 hours. Seeding was
performed semidynamically by incubation of the scaffolds overnight in a cell suspension with
gentle shaking. Osteogenic differentiation of hMSC was induced by addition of 10^{-7} M dexamethasone, 3.5 mM β-glycerophosphate, and 0.05 mM ascorbic acid 2-phosphate to the cell
culture medium. Cell morphology and distribution was observed by SEM on (a, b) day 6 and
(c, d) day 12. Proliferation and osteogenic differentiation were detected by measurement of (e)
cytosolic LDH activity and (f) specific ALP activity.

collagen type I). Due to its composite character, bone is perfectly adapted to compression, tension, bending, and torsion.

The following sections describe two different types of CPC–alginate composite scaffolds. Beside monophasic composite scaffolds prepared of a CPC–alginate mixture paste, biphasic structures are fabricated through simultaneous plotting of a CPC and an alginate paste demonstrating the option to generate biphasic organic and inorganic scaffolds with a designed structure adapted to special requirements of complex tissue defects.

4.3.1 Fabrication of Biphasic and Mixed CPC–Alginate Scaffolds

Alginate is a polysaccharide mostly isolated from brown algae. Its specific properties make it interesting for biomedical applications: By dissolving sodium alginate powder in water or aqueous solutions, an alginate sol can be prepared whose transition into a gel is induced by multivalent ions such as Ca^{2+}, that is, under mild conditions. Alginate was selected as a polymer phase for preparation of the composite scaffolds due to the favorable characteristics of biocompatibility and biodegradability as well as the simple and nontoxic cross-linking process, which can be performed simultaneously to the CPC setting reaction.

In a first step, an alginate paste was developed whose consistency allows its plotting in combination with the P-CPC (Section 4.2). A highly concentrated alginate–polyvinyl alcohol (PVA) paste with an alginate content of 16.7 wt% was prepared by mixing alginate powder with 6% w/v PVA solution in a mass ratio of 1 to 5 until a homogeneous plotting paste was achieved. The P-CPC and the alginate–PVA paste were transferred into two different cartridges and used for fabrication of biphasic CPC–alginate scaffolds by alternate extrusion of CPC and alginate strands (Figure 4.10, left).

Compared to low concentrated alginate sols (1–4 wt%), which are normally used for the preparation of hydrogels and scaffolds, the highly concentrated alginate–PVA paste has significant advantages for 3D plotting of scaffolds (Luo, Wu, et al. 2013). First, its consistency is comparable to those of the P-CPC enabling simultaneous processing of both pastes in one and the same scaffold. This alginate–PVA paste can support the whole structure of the printed scaffold without deformation. Second, in contrast to the less viscous low concentrated alginate sols, the viscosity of the highly concentrated alginate–PVA paste is high enough to allow plotting in air instead of a $CaCl_2$ solution as it would be necessary otherwise to stabilize the structures immediately after extrusion. This is an important prerequisite for plotting in combination with the P-CPC and can avoid problems, caused by extrusion into a $CaCl_2$ bath such as blocking of the nozzle during the manufacturing process and inhomogeneous cross-linking of the scaffold with incomplete or no bonding between the strands.

For fabrication of the monophasic composite scaffolds, a mixture paste consisting of CPC precursor powder (Khairoun et al. 1997) and alginate was

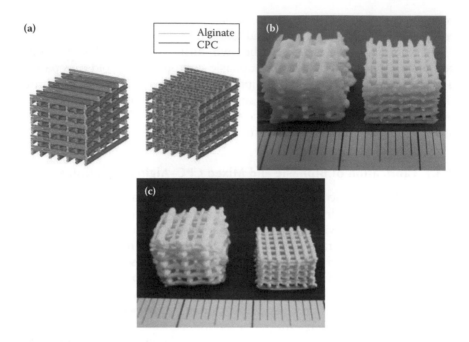

FIGURE 4.10
(See color insert.) Biphasic and mixed CPC–alginate scaffolds. (a) CAD models and realized
scaffolds in (b) wet and (c) dry state are shown. The biphasic CPC–alginate scaffold (left) was
produced by alternate plotting of P-CPC and a high-concentrated alginate–PVA paste through
nozzles with an inner diameter of 610 μm (printing speed: 3 mm/s, dosing air pressure: 5.0
bar for the P-CPC and 7.4 bar for the alginate–PVA paste). For fabrication of the mixed CPC–
alginate scaffold (right), a CPC–alginate–PVA paste was plotted through a nozzle with inner
diameter of 610 μm (printing speed: 3 mm/s, dosing air pressure: 8.5 bar).

applied. To obtain a paste suitable for plotting of such scaffolds, CPC powder
and alginate powder (in a mass ratio of 1 to 1) were mixed first, and then 1
g of the mixed CPC–alginate powder was added to 4.5 g of a 6% w/v PVA
solution, stirred until a homogeneous paste was achieved and filled in a cartridge of the printing head.

Both types of composite scaffolds, the biphasic CPC–alginate ones and the
mixed CPC–alginate scaffolds, were fabricated with a 0°0°/90°90° (XXYY)
lay-down pattern: two layers of the paste were deposited in the X direction
on top of each other and then in the Y direction the next two layers; this
sequence was repeated until the whole scaffold was finished. The application of the XXYY lay-down pattern enhances the pore size in the x- and
y-direction and improves the interconnectivity of the pores compared to
the conventional XY lay-down pattern (i.e., alternating deposition of each
layer). After completion of the plotting process the biphasic CPC–alginate
scaffolds as well as the mixed CPC–alginate scaffolds were incubated in a 1
M CaCl₂ solution for cross-linking of the alginate and simultaneous setting
of the P-CPC for 5 h. Afterward, the CPC–alginate biphasic scaffolds were

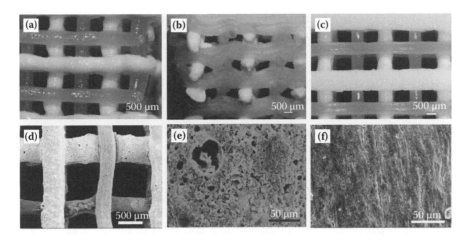

FIGURE 4.11
Microscopic images of plotted biphasic CPC–alginate scaffolds: (a) Top, (b) side, and (c) bottom view and (d) SEM images of the top view as well as high-magnification scanning electron micrographs of (e) CPC and (f) alginate strands.

transferred into deionized water and incubated for 1 week with a change of water twice per day to accomplish hardening of the CPC and to remove the incorporated oil from the P-CPC phase. PVA in the alginate strands was mostly dissolved during this time. The mixed CaP–alginate scaffolds were washed three times with deionized water. Finally, the composite scaffolds were dried at air and room temperature.

In Figure 4.10, the CAD models of both composite scaffold types and the real objects achieved by the 3D plotting process are shown. The photographs demonstrate that the designed biphasic CPC–alginate scaffold with P-CPC and alginate strands plotted alternately was successfully realized.

The strands of the different layers were bond closely among one another due to the homogeneous cross-linking of alginate and setting of CPC. The biphasic scaffolds were further stabilized by interpenetration of the cross-linked alginate and hardened CPC strands between the layers. Micrographs show the regular structure and organization of CPC and alginate strands that are alternatively weaved forming an interpenetrating structure consisting of an alginate and a CPC grid with macropores, open in all directions (Figure 4.11). High magnification SEM images revealed that the CPC strands possessed an additional microporosity and alginate strands were dense and smooth (Figure 4.11). Mixed CPC–alginate scaffolds with completely open macropores were also successfully fabricated using the same design parameters. In this composite type of scaffold, the CPC particles were evenly distributed within the alginate strands.

As it is known that CPC sets and hardens without any dimensional change (Fernandez et al. 1995), no significant change of size was recognized for the pure CPC scaffolds after drying. In contrast, pure alginate and mixed

CPC–alginate scaffolds underwent shrinkage in the x-, y-, and z-directions to a different extent during drying due to the loss of water from alginate pastes (Figure 4.10). Scaffolds printed from highly concentrated alginate also undergo shrinkage to a certain degree but without deformation, in contrast to scaffolds prepared from low-concentrated alginate sols that suffer from severe shrinkage and deformation during drying. In case of the biphasic CPC–alginate scaffolds, only the alginate strands shrunk and were thereby stretched along the plotting direction; however, almost no shrinkage was observed for the whole scaffold due to the support of CPC strands.

4.3.2 Characterization of the Plotted CPC–Alginate Scaffolds

To evaluate the potential of the novel biphasic CPC–alginate scaffolds, their mechanical properties, cytocompatibility, and protein delivery ability were studied *in vitro* in comparison to scaffolds plotted of pure P-CPC, pure 16.7 wt% alginate paste, and the mixed CPC–alginate paste.

The compressive strength and modulus of the plotted scaffolds in dry and wet state (after incubation in simulated body fluid [SBF] for 2 h) were tested by using an Instron 5566 device with a load cell of 10 kN and a constant rate of 1 mm/min. The obtained results including compressive strength, modulus, and stress–strain curves of the scaffolds are presented in Figure 4.12. The compressive strength of both the biphasic CPC–alginate scaffolds and the mixed CPC–alginate scaffolds was almost twice as high as that of pure CPC ones in dry state. In wet state, the compressive strength of the pure alginate and the mixed CPC/alginate scaffolds decreased sharply to values that are significantly lower than that of biphasic CPC–alginate and pure CPC scaffolds. The modulus of CPC and biphasic CPC–alginate were higher compared to that of alginate and mixed CPC–alginate scaffolds in the wet state but showed no significant difference in the dry state. The most likely explanation is the decrease of stiffness of alginate gels due to swelling in aqueous media. Interestingly, compressive strength of biphasic CPC–alginate scaffolds was still significantly higher than that of pure CPC scaffolds in the wet state indicating the beneficial effect of the combination of both phases. Figure 4.12c,d show that the stress of pure alginate and mixed CPC–alginate scaffolds increased with compressive deformation and the scaffolds still held the bulk morphology without collapsing. Stress of pure CPC scaffold increased sharply with compression at the beginning and then fell quickly to zero after reaching the maximum value as the scaffold completely disintegrated (Figure 4.12c,d). In contrast, the compressive strength of the biphasic CPC–alginate scaffolds increased almost linearly with deformation at the beginning and decreased a little after the maximum value. After 35% compressive deformation was performed, the biphasic CPC–alginate scaffolds still maintained their macroscopic morphology instead of cracking into powder as observed for pure CPC ones (Figure 4.12d).

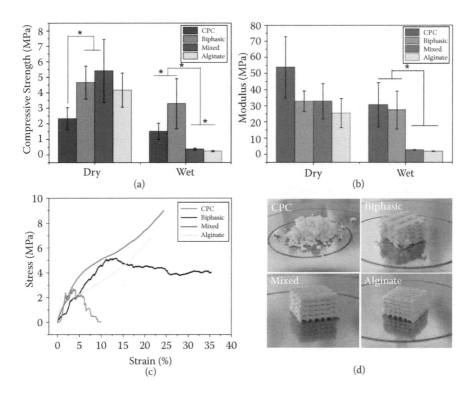

FIGURE 4.12

(a) Compressive strength and (b) modulus of plotted CPC–alginate scaffolds in dry and wet state, and (c) compression curves and (d) photographs of dry scaffolds after compression of up to 30% deformation. All scaffolds were prepared with the same geometry (strand distance and number of strands). Compressive strength and modulus of pure alginate and alginate–CPC mixed scaffolds were calculated on the values at 15% strain, while for the CPC and CPC–alginate biphasic scaffolds by using the maximal values.

Whereas pure CPC scaffolds, fabricated without sintering, are characterized by their inherent brittleness and poor mechanical strength, most pure polymer scaffolds possess an insufficient stiffness. Pure alginate and mixed CPC–alginate scaffolds are polymeric bulk materials, which show the typical behavior of stress increase with growing compressive deformation but without collapse of the structures. The pure CPC scaffolds, on the other hand, react as expected for ceramic bulk materials with a sharp increase of stress at the beginning of the compression followed by a drop to nearly zero quickly after reaching the maximum value, that is caused by crack formation, resulting in a complete collapse of the architecture. Biphasic CPC–alginate scaffolds consist of both polymeric and ceramic bulk material. In case of these composite scaffolds, stress increased with the compressive strain linearly at the beginning until the maximum, which was significantly higher than those of pure CPC scaffolds. In contrast to the latter, the stress did not fall to

zero but maintained a certain strength due to the contribution of supporting alginate. Compressive strength of the biphasic scaffolds was therefore
enhanced compared to that of pure CPC ones with identical porosity and
geometry due to the binding effect of the alginate strands. Furthermore, the
alginate strands formed an interpenetrating network that further stabilized
the CPC strands in these biphasic scaffolds. That is why the biphasic scaffolds
were able to hold their bulk morphology after suffering as much as 30% compressive strain. Compared to pure CPC scaffolds, the biphasic ones showed
improved toughness. The high compressive strength of pure alginate scaffolds in dry conditions can be explained by the apparent shrinkage of these
scaffolds during dehydration, resulting in a lower porosity, higher solid state
content, and stronger entanglement of the biopolymer chains. Therefore, in
a wet state the strength of pure alginate and mixed alginate–CPC scaffolds
decreased sharply due to the swelling of alginate hydrogel strands.

Cell culture experiments performed *in vitro* are able to provide information concerning cytocompatibility being the first step of evaluating the performance of a novel type of scaffold for tissue engineering and regenerative
therapies. As already demonstrated, CPC scaffolds are good candidates supporting the attachment, growth, and osteogenic differentiation of hMSC.
Cytocompatibility of both the biphasic and mixed CPC–alginate composite
scaffolds in comparison to the pure CPC ones was evaluated by seeding of
hMSC on the printed structures. The cell-seeded scaffolds were cultured
over a period of 3 weeks with or without osteogenic supplements. The morphology and distribution of the cells attached to the scaffolds were observed
by means of SEM: one day after seeding, hMSC were attached and well
spread on the surface of the biphasic CPC–alginate as well as mixed scaffolds
(Figure 4.13). Interestingly, in the course of further cultivation, differences
have been found: while viable and well spread cells were detected on pure
CPC and biphasic CPC–alginate scaffolds, both in the presence and in the
absence of osteogenic supplements, the number of cells cultivated on mixed
CPC–alginate scaffolds was strongly reduced but only in the presence of
osteogenic supplements, as indicated by SEM analysis and determination of
cytosolic LDH activity (as measured for the number of living cells). This phenomenon might be caused by the apatite formation on the surface of mixed
CPC–alginate scaffolds that was observed to start on day 1 of culture resulting in an intense deposition of apatite particles as observed after 7 days of
culture (Figure 4.13d). A potential reason for the strong apatite precipitation
on the mixed CPC–alginate scaffolds was the high degree of supersaturation
due to the slow hydrolysis of CPC particles (mostly α-TCP) and both calcium ion exchange from the Ca^{2+}-cross-linked alginate and the elevated local
phosphate ion concentrations because of the presence of β-glycerophosphate
at the same time (Erisken et al. 2011).

Scaffolds with the capability to bind growth factors (GF) and drugs, and
release them in a controlled manner are preferable candidates for tissue

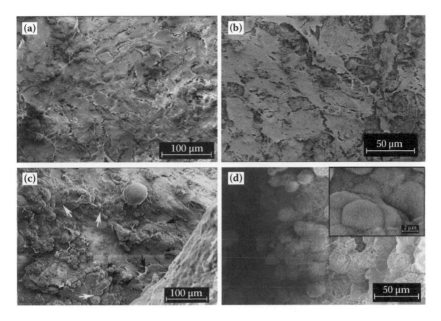

FIGURE 4.13
SEM images of hMSC attached on the surface of (a, b) biphasic CPC–alginate scaffolds and (c, d) mixed CaP–alginate scaffolds on (a, c) day 1 and (b, d) day 7 of culture in the presence of osteogenic supplements (white arrows indicate the formed apatite deposition and black arrow attached cells).

engineering and regeneration. The incorporation of GF and drugs into a biomaterial prior to scaffold building is a good strategy to heighten loading efficiency and to prolong the release period in comparison to surface-loading approaches. However, most methods of scaffold fabrication involve usage of organic solvents or heat treatment making the incorporation of biological components into the material impossible, in contrast to the 3D plotting technique applied in our work for building of biphasic and mixed alginate–CPC scaffolds that took place under nearly physiological conditions. The integration of proteins into scaffolds during the plotting process was evaluated by mixing of bovine serum albumin (BSA), as a model protein, into the pastes, which were used to produce pure alginate, pure CPC, and mixed alginate–CPC scaffolds. BSA was homogeneously distributed in the plotted structures without denaturation. The initial burst of release within the first 12 h was determined to be lower than 20% for CPC and 40% for alginate scaffolds. For the mixed alginate–CPC scaffolds intermediate amounts were released within the first 12 h. Later on, a sustained release behavior has been observed for all three scaffold types but with considerable differences with respect to the amount of released BSA. The pure CPC scaffolds released clearly smaller amounts of BSA signifying that the main part of BSA remained bound to the calcium phosphate matrix. In contrast, much higher amounts of BSA were

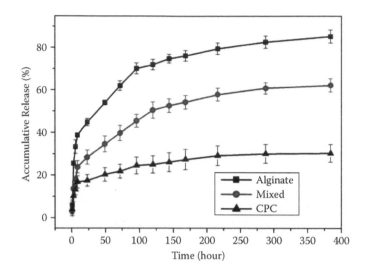

FIGURE 4.14
BSA release from plotted scaffolds in SBF at 37°C over a period of 16 days.

released from the pure alginate scaffolds because of the high nanoporosity of the hydrogel and quick degradation of alginate scaffolds as well as the low protein binding capacity of alginate in general. Again, for the mixed alginate–CPC scaffolds an intermediate release profile was detected. It can be speculated that the release of proteins from biphasic CPC–alginate scaffolds can be controlled by altering the loading amount of BSA in CPC and alginate strands (Figure 4.14).

Our data demonstrate the beneficial effect of the combination of ceramics and polymers by alternate extrusion that result in biphasic scaffolds. Especially with respect to the mechanical properties the performance of the biphasic CPC–alginate composite scaffolds exceed that of those fabricated by 3D plotting of a mixed paste of CPC and alginate. In addition, the biphasic CPC–alginate scaffolds performed better in first cell culture experiments compared to the mixed ones.

Furthermore, by using the 3D plotting technique and a device that allows the utilization of more than one pasty material for one scaffold, an upgraded biphasic scaffold with separate layers suitable for repair of defects at tissue interfaces can be developed (Figure 4.15). Such a complex scaffold could be designed, for example, for an osteochondral defect, by combining an organic/inorganic composite layer to fill the bony part of the lesion and a non-mineralized organic layer for the repair of the chondral part. To ensure that both layers are tightly connected to each other, they have to be plotted simultaneously and some of the strands of the organic phase have to penetrate the other (composite) layers realizing a mechanically stable interlocking of both materials.

FIGURE 4.15
Photograph of a biphasic scaffold with separate layers for osteochondral tissue engineering. The bottom layers were plotted from a CPC–alginate mixture for treatment of the bony part and the top layers consist of alginate for the chondral part of the defect.

4.4 3D Plotting of Silicon-Based Bioceramic Scaffolds

Silicon-based bioceramics and scaffolds play an increasing role in bone tissue engineering because silicon is one of the important trace elements in human bone and is reported to stimulate new bone growth (Mertz 1981; Pietak et al. 2007). Conventional methods for fabrication of porous silicon-based bioceramic scaffolds are particle leaching and polyurethane template sintering (Wu et al. 2010). As previously discussed, the drawbacks of these methods do not only include poor control of pore parameters but also weak mechanical properties of these scaffolds. Therefore, our intention was to fabricate silicon-based scaffolds with designed pores with the application of rapid prototyping.

In our work, two types of silicon-based scaffolds were printed by using pastes containing either β-CaSiO$_3$ or mesoporous Ca-Si-P bioglass. β-CaSiO$_3$ is one of the most important silicon compounds used as bone replacement material because β-CaSiO$_3$ degrades faster and has been demonstrated to stimulate bone formation *in vivo* better than β-tricalcium phosphate (β-TCP) (Xu et al. 2008). Ca-Si-P bioglass with regular nanochannels and high specific surface area is a bioactive material. Numerous reports have demonstrated that Ca-Si-P bioglass not only is an excellent drug delivery system but also a potential candidate for bone repair (López-Noriega et al. 2006). β-CaSiO$_3$ powder was synthesized by a chemical precipitation method (Wu et al. 2012). Ca-Si-P bioglass was synthesized by a sol-gel and assembly technique according to a published protocol (Wu et al. 2010). The product was ground and passed through 300 meshes. Using this method, regular nanochannels with diameters of around 5 nm were formed in the Ca-Si-P bioglass particles, which enhance the specific surface area making this material attractive for drug-loading applications by increasing the binding capability and sustained release ability (Xia and Chang 2006).

For preparation of the plotting pastes, β-CaSiO$_3$ and Ca-Si-P bioglass powder were further sieved to reduce the particle size to less than 45 μm and mixed with a 15 wt% aqueous PVA solution. As previously described, the biocompatible polymer PVA is ideally suited to prepare bioceramic pastes with excellent injectability. After stirring the mixtures (with adjusted powder-to-liquid ratio) to become homogeneous pastes and loading into a printing cartridge, the plotting process was carried out as described earlier. Plotted scaffolds were dried at 40°C overnight, and then heated to 150°C for 30 min to induce heat-activated cross-linking of PVA. Afterward, the obtained scaffolds were stable in dry state as well as in aqueous solutions like cell culture medium without a second high-temperature sintering step. In this system, PVA works as a binder to bond the bioceramic particles. The molecular weight of PVA was 130,000 and after cross-linking, the crystallinity of PVA was increased and solubility therefore diminished (Chun et al. 2010). A PVA content of only 12 wt% for plotted CaSiO$_3$ and 14 wt% for plotted mesoporous Ca-Si-P scaffolds in dry state was enough to bind the ceramic powders.

In further studies we have revealed that the 3D plotted silicon-based bioceramic scaffolds with PVA as binder had excellent apatite deposition ability after incubation in SBF *in vitro* and new bone formation ability *in vivo* (Wu et al. 2012). No negative effect of PVA on the bioactivity has been observed. The pore size and morphology of the two types of plotted silicon-based bioceramic scaffolds were characterized microscopically (Figure 4.16). With the control of CAD, scaffolds with defined geometry and porosity were prepared of both pastes. However, concerning the surface structure, certain differences are visible: the surface of the CaSiO$_3$ scaffold was rough and porous, and that of the mesoporous bioglass scaffolds was smooth and dense. Those differences were related to the powder particle size (β-CaSiO$_3$ particles were bigger than those of mesoporous Ca-Si-P) and the higher specific surface area of the Ca-Si-P scaffolds was due to the mesoporosity of the bioglass.

The compressive strength of plotted scaffolds was tested by exerting the force in z-direction using an Instron 5566 testing machine equipped with a load cell of 10 kN at a crosshead speed of 0.5 mm/min. The stress–strain curves and morphologies of scaffolds before and after compression are shown in Figure 4.17. The maximal compressive strength and modulus of plotted β-CaSiO$_3$ scaffolds with a porosity of 65% were 3.6±0.1 and 39.5±7.7 MPa, respectively. Compressive strength and modulus of plotted mesoporous Ca-Si-P scaffolds with a porosity of 60.4% were 16.1±1.5 and 155.1±14.9 MPa, respectively. The compressive strength and modulus of these scaffolds were significantly higher than that of silicon-based bioceramic scaffolds of similar porosity, prepared by conventional methods, especially for the

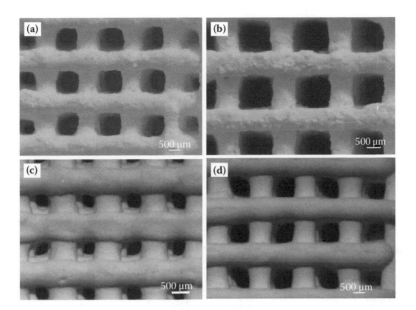

FIGURE 4.16
Light micrographs of plotted (a, b) β-CaSiO$_3$ and (c, d) mesoporous Ca-Si-P scaffolds with a different pore size.

mesoporous Ca-Si-P scaffolds. For example, for mesoporous bioglass scaffolds prepared by polyurethane templating a compressive strength of only 0.08 MPa was reported (Wu et al. 2011), which is 200 times lower than that of the plotted mesoporous Ca-Si-P scaffolds described herein. In addition, compared to other polymer–ceramics composite scaffolds, the PVA–Ca-Si-P composite scaffolds prepared in this work also have significantly stronger compressive strength. For example, Kalita et al. (2003) fabricated polypropylene–TCP composite scaffolds by fused deposition modeling, leading to similar structures than those achieved by 3D plotting. The compressive strength of the polypropylene–TCP composite scaffolds with a porosity of 36% was 12.7 MPa (±2 MPa), which is significantly lower than that of our PVA/Ca-Si-P composite scaffolds, even though these had a much higher porosity of 60%. We found that the compressive strength and modulus of the plotted mesoporous Ca-Si-P scaffolds were clearly higher compared to that of the plotted β-CaSiO$_3$ scaffolds. The potential reason for this observation relates on the structure of the mesoporous Ca-Si-P particles with their regular nanochannels and higher specific surface area: the Ca-Si-P particles can be bonded stronger by PVA due to a partial filling of the nanochannels with the polymer resulting in denser strands after extrusion.

The stress–strain curves of both bioceramic scaffold types indicate that the stress increased linearly with increasing compressive strain until about 10% deformation (Figure 4.17a,b). After the maximum stress was achieved, the compressive stress decreased with further increasing strain. However,

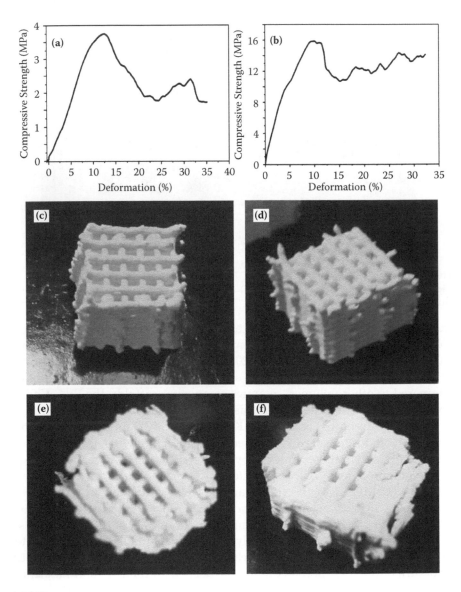

FIGURE 4.17
Compressive strength of plotted silicon-based bioceramic scaffolds. Stress–strain curves of plotted (a) β-CaSiO$_3$ and (b) mesoporous Ca-Si-P scaffolds; (c, e) morphologies of β-CaSiO$_3$ and (d, f) mesoporous Ca-Si-P scaffolds (c, d) before and (e, f) after compression.

in contrast to the behavior of conventional pure ceramic scaffolds characterized by a decline of stress to zero after the maximum, the compressive stress of plotted silicon-based scaffolds with PVA as binder still hold certain strength after reaching the maximum. For example, the plotted β-CaSiO$_3$ and mesoporous Ca-Si-P scaffolds still possessed about 43% and 85% of their maximum stress at 35% of strain, respectively. The photographs taken of the plotted scaffolds before and after mechanical testing, respectively, show that after 35% of compression the printed silicon-based scaffolds still retained their integrity. Porous ceramic scaffolds fabricated by conventional methods such as polyurethane templating are brittle and very easy to disintegrate to powder after compression, which is the main drawback of these bioceramics. However, the silicon-based scaffolds prepared by 3D plotting with PVA as polymeric binder do not only have increased mechanical stability but also improved toughness. As a macromolecule, PVA strongly bonds the ceramic particles after cross-linking. After compression testing, PVA fibers binding the particles were clearly visible in SEM images (Wu et al. 2011).

The composite pastes prepared by mixing bioceramics (such as silicon-based ceramics) and PVA in certain mass ratio were injectable and suitable for extrusion through a nozzle with inner diameter of 610 μm to form 3D structures of defined inner and outer morphology by 3D plotting. The successful and advantageous application of the approach of 3D plotting of bioceramic particles by using PVA as a polymeric binder might not be limited to the silicon-based bioceramics described herein but also useful for fabrication of other kinds of bioceramic scaffolds consisting of mineral phases like HA or β-TCP.

4.5 Conclusions and Future Directions

In this chapter, bioceramic scaffolds with predefined inner and outer morphology, fabricated by 3D plotting of a calcium phosphate cement (CPC) and silicon-based ceramics, respectively, were described. The newly prepared CPC pastes allowed extended plotting and set under physiological conditions. In addition, the mechanical properties including compressive strength and toughness of plotted CPC scaffolds were improved by simultaneous plotting of concentrated alginate and CPC pastes, leading to biphasic structures. Such biphasic scaffolds were also favorable for cell attachment as well as able to maintain controlled protein delivery. Furthermore, another biphasic scaffold with separate layers for bone and cartilage was designed for osteochondral tissue engineering and realized by plotting a CPC–alginate mixture for the bony and alginate for the chondral part. On the other hand, two novel silicon-based ceramic scaffolds (CaSiO$_3$ and mesoporous bioglass) fabricated by 3D plotting were also introduced. These materials with tailorable structures showed improved mechanical properties due to the utilization of PVA as binder and the regular orientation of pores.

Rapid prototyping technologies like 3D plotting have been demonstrated to be useful methods for preparation of adjustable scaffolds (made of polymers, ceramics, and composites) for tissue engineering applications. However, most of the work performed in this field so far follows a two-step strategy of first manufacturing the scaffold and afterward seeding it with cells or loading it with additional biological components. Three-dimensional plotting of pasty biomaterials that form stable structures under physiological or at least mild conditions (like the ones described here) opens the possibility to include biological components directly in the fabrication process. This approach facilitates the seeding of constructs with more than one cell type, which definitely would be necessary for the engineering of complex tissues and vascularized structures. We are therefore sure that further development of 3D plotting including biological components will lead to relevant new findings and finally to clinically applicable solutions.

Acknowledgments

Most of the work presented in this chapter was carried out at the Centre for Translational Bone, Joint and Soft Tissue Research, University Hospital Carl Gustav Carus and Medical Faculty of Technische Universität Dresden, Germany. The authors thank Dr. Berthold Nies (InnoTERE GmbH, Dresden, Germany) for providing the CPC pastes; and Dr. Frank Sonntag and his colleagues Katrin Meissner and Jens Steidl (Fraunhofer IWS Dresden, Germany) for developing and upgrading the plotting system. We also thank Professor Martin Bornhäuser and coworkers (Medical Clinic I, Dresden University Hospital Carl Gustav Carus) for providing hMSC. This research was funded by German Federal Ministry for Economy and Technology (BMWi) via AIF (project No. IGF 306 ZBR).

References

Bohner, M. 2000. Calcium orthophosphates in medicine: From ceramics to calcium phosphate cements. *Injury* 31:37–47.

Chun, J. Y., Kang, H. K., Jeong, L., et al. 2010. Epidermal cellular response to poly (vinyl alcohol) nanofibers containing silver nanoparticles. *Colloids Surf B Biointerfaces* 78:334–42.

Constantz, B. R., Ison, I. C., Fulmer, M. T., et al. 1995. Skeletal repair by in situ formation of the mineral phase of bone. *Science* 267:1796–9.

Detsch, R., Uhl, F., Deisinger, U., and Ziegler, G. 2008. 3D-Cultivation of bone marrow stromal cells on hydroxyapatite scaffolds fabricated by dispense-plotting and negative mould technique. *J Mater Sci: Mater Med* 19:1491–6.

Dorozhkin, S.V. 2010. Bioceramics of calcium orthophosphates. *Biomaterials* 31:1465–85.

Erisken, C., Kalyon, D., Wang, H., Örnek-Balance, C., and Xu, J. 2011. Osteochondral tissue formation through adipose-derived stromal cell differentiation on biomimetic polycaprolactone nanofibrous scaffolds with graded insulin and beta-glycerophosphate concentrations. *Tissue Eng A* 17:1239–52.

Espanol, M., Perez, R. A., Montufar, E. B., Marichal, C., Sacco, A., and Ginebra, M. P. 2009. Intrinsic porosity of calcium phosphate cements and its significance for drug delivery and tissue engineering applications. *Acta Biomater* 5:2752–62.

Fedorovich, N. E., De Wijn, J. R., Verbout, A. J., Alblas, J., and Dhert, W. J. A. 2008. Three-dimensional fiber deposition of cell-laden, viable, patterned constructs for bone tissue printing. *Tissue Eng Part A* 14:127–33.

Fernandez, E., Ginebra, M. P., Bermodez, O., Boltong, M. G., Driessens, F. C. M., and Planell, J. A. 1995. Dimensional and thermal behaviour of calcium phosphate cements during setting compared to PMMA bone cements. *J Mater Sci Lett* 14:4–5.

Franco, J., Hunger, P., Launey, M. E., Tomsia, A. P., and Saiz, E. 2010. Direct write assembly of calcium phosphate scaffolds using a water-based hydrogel. *Acta Biomater* 6:218–28.

Gelinsky, M. 2009. Mineralised collagen as biomaterial and matrix for bone tissue engineering. In U. Meyer, T. Meyer, J. Handschel, and H. P. Wiesmann (eds.), *Fundamentals of Tissue Engineering and Regenerative Medicine*, 485–493. Springer, Heidelberg and Berlin.

Gelinsky, M., and Heinemann, S. 2010. Nanocomposites for tissue engineering. In C. Kumar (ed.), *Nanocomposites for Life Sciences*, Chapter 11, 405–34. Wiley-VCH, Weinheim.

Gelinsky, M., Lode, A., Bernhardt, A., and Rösen-Wolff, A. 2011. Stem cell engineering for regeneration of bone tissue. In G. M. Artmann, J. Hescheler, and S. Minger (eds.), *Stem Cell Engineering*, Part 3, 383–99. Springer, Heidelberg and Berlin.

Guo, H., Su, J., Wei, J., Kong, H., and Liu, C. 2009. Biocompatibility and osteogenicity of degradable Ca-deficient hydroxyapatite scaffolds from calcium phosphate cement for bone tissue engineering. *Acta Biomater* 5:268–78.

Hutmacher, D. W., Schantz, T., Zein, I., Ng, K. W., Teoh, S. H., and Tan, K. C. 2001. Mechanical properties and cell cultural response of polycaprolactone scaffolds designed and fabricated via fused deposition modeling. *J Biomed Mater Res* 55:203–16.

Hutmacher, D. W., Sittinger, M., and Risbud, M. V. 2004. Scaffold-based tissue engineering: Rationale for computer-aided design and solid free-form fabrication systems. *Trends Biotechnol* 22:354–62.

Jansen, J. A., de Ruijter, J. E., Schaeken, H. G., van der Waerden, J. P. C. M., Planell, J. A., and Driessens, F. C. M. 1995. Evaluation of tricalcium-phosphate/hydroxyapatite cement for tooth replacement: An experimental animal study. *J Mater Sci Mater Med* 6:653–7.

Jones, A. C., Arns, C. H., Hutmacher, D. W., Milthorpe, B. K., Sheppard, A. P., and Knackstedt, M. A. 2009. The correlation of pore morphology, interconnectivity and physical properties of 3D ceramics scaffolds with bone ingrowth. *Biomaterials* 30:1440–51.

Kalita, S. J., Bose, S., Hosick, H. L., and Bandyopadhyay, A. 2003. Development of controlled porosity polymer-ceramic composite scaffolds via fused deposition modeling. *Mater Sci Eng C* 23:611–620.

Karageorgiou, V., and Kaplan, D. 2005. Porosity of 3D biomaterial scaffolds and osteogenesis. *Biomaterials* 26:5474–91.

Keating, J. F., and McQueen, M. M. 2001. Substitutes for autologous bone graft in orthopaedic trauma. *J Bone Joint Surg Br* **83**:3–8.

Khairoun, I., Boltong, M. G., Driessens, F. C., and Planell, J. A. 1997. Effect of calcium carbonate on clinical compliance of apatitic calcium phosphate bone cement. *J Biomed Mater Res* 38:356–60.

Khalyfa, A., Vogt, S., Weisser, J., et al. 2007. Development of a new calcium phosphate powder-binder system for the 3D printing of patient specific implants. *J Mater Sci Mater Med* 18:909–16.

Kim, G. H., and Son, J. G. 2009. 3D polycaprolactone (PCL) scaffold with hierarchical structure fabricated by a piezoelectric transducer (PZT)-assisted bioplotter. *Appl Phys A* 94:781–5.

Landers, R., Hübner, U., Schmelzeisen, R., and Mülhaupt, R. 2002. Rapid prototyping of scaffolds derived from thermoreversible hydrogels and tailored for applications in tissue engineering. *Biomaterials* 23:4437–47.

Leong, K. F., Chua, C. K., Sudarmadji, N., and Yeong, W. Y. 2008. Engineering functionally graded tissue engineering scaffolds. *J Mech Behav Biomed Mater* 1:140–52.

Lode, A., Meissner, K., Luo, Y., et al. Forthcoming. Fabrication of porous scaffolds by three-dimensional plotting of a pasty calcium phosphate bone cement under mild conditions. *J Tissue Eng Regen Med*. DOI 10.1002/term.1563.

López-Noriega, A., Arcos, D., Izquierdo-Barba, I., Sakamoto, Y., Terasaki, O., and Vallet-Regí, M. 2006. Ordered mesoporous bioactive glasses for bone tissue regeneration. *Chem Mater* 18:3137–44.

Luo, Y., Lode, A., and Gelinsky, M. Forthcoming. Direct plotting of three-dimensional hollow fiber scaffolds based on concentrated alginate pastes for tissue engineering. *Adv Healthcare Mater*. DOI 10.1002/adhm.201200303.

Luo, Y., Wu, C., Lode, A., and Gelinsky, M. Forthcoming. Hierarchical mesoporous bioactive glass/alginate composite scaffolds fabricated by three-dimensional plotting for bone tissue engineering. *Biofabrication*. 5:015005.

Maher, P. S., Keatch, R. P., Donnelly, K., and Mackay, R. E. 2009. Construction of 3D biological matrices using rapid prototyping technology. *Rapid Prototyp J* 15:204–10.

Martins, A., Chung, S., Pedro, A. J., Sousa, R. A., Marques, A. P., Reis, R. L., and Neves, N. M. 2009. Hierarchical starch-based fibrous scaffold for bone tissue engineering applications. *J Tissue Eng Regen Med* 3:37–42.

Mertz, W. 1981. The essential trace elements. *Science* 213:1332–8.

Miranda, P., Pajares, A., Saiz, E., Tomsia, A. P., and Guiberteau, F. 2007. Fracture modes under uniaxial compression in hydroxyapatite scaffolds fabricated by robocasting. *J Biomed Mater Res A* 83:646–55.

Miranda, P., Pajares, A., Saiz, E., Tomsia, A. P., and Guiberteau, F. 2008. Mechanical properties of calcium phosphate scaffolds fabricated by robocasting. *J Biomed Mater Res A* 85:218–27.

Miranda, P., Saiz, E., Gryn, K., and Tomsia, A. P. 2006. Sintering and robocasting of β-tricalcium phosphate scaffolds for orthopaedic applications. *Acta Biomater* 2:457–66.

Mironov, V., Visconti, R. P., Kasyanov, V., Forgacs, G., Drake, C. J., and Markwald, R. R. 2009. Organ printing: Tissue spheroids as building blocks. *Biomaterials* 30:2164–74.

Murphy, C. M., Haugh, M. G., and O'Brien, F. J. 2010. The effect of mean pore size on cell attachment, proliferation and migration in collagen–glycosaminoglycan scaffolds for bone tissue engineering. *Biomaterials* 31:461–6.

Nemzek, J. A., Arnoczky, S. P., and Swenson, C. L. 1994. Retrovinal transmission by the transplantation of connective-tissue allografts. An experimental study. *J Bone Joint Surg Br* 76:1036–41.

O'Brien, F. J. 2011. Biomaterials and scaffolds for tissue engineering. *Mater Today* 14:88–95.

Oliveira, A. L., Costa, S. A., Sousa, R. A., and Reis, R. L. 2009. Nucleation and growth of biomimetic apatite layers on 3D plotted biodegradable polymeric scaffolds: Effect of static and dynamic coating conditions. *Acta Biomat* 5:1626–38.

Paul, J., Stagstetter, A., Kriner, M., et al. 2009. Donor-site morbidity after osteochondral autologous transplantation for lesions of the talus. *J Bone Joint Surg Br* 91:1683–8.

Pfister, A., Landers, R., Laib, A., Huebner, U., Schmelzeisen, R., and Muelhaupt, R. 2004. Biofunctional rapid prototyping for tissue-engineering applications: 3D bioplotting versus 3D printing. *J Polym Sci Part A: Polym Chem* 42:624–38.

Pietak, A. M., Reid, J. W., Stott, M. J., and Sayer, M. 2007. Silicon substitution in the calcium phosphate bioceramics. *Biomaterials* 28:4023–32.

Sachs, E., Cima, M., Cornie, J., et al. 1993. Three-dimensional printing: The physics and implications of additive manufacturing. *CIRP Ann Manufac Technol* 42:257–60.

Schroeder, J. E., and Mosheiff, R. 2011. Tissue engineering approaches for bone repair: Concepts and evidence. *Injury* 42:609–13.

Seitz, H., Rieder, W., Irsen, S., Leukers, B., and Tille, C. 2005. Three-dimensional printing of porous ceramic scaffolds for bone tissue engineering. *J Biomed Mater Res B* 74:782–8.

Sobral, J. M., Caridade, S. G., Sousa, R. A., Mano, J. F., and Reis, R. L. 2011. Three-dimensional plotted scaffolds with controlled pore size gradients: Effect of scaffold geometry on mechanical performance and cell seeding efficiency. *Acta Biomater* 7:1009–18.

Srouji, S., Kizhner, T., Suss-Tobi, E., Livne, E., and Zussman, E. 2008. 3-D Nanofibrous electrospun multilayered construct is an alternative ECM mimicking scaffold. *J Mater Sci: Mater Med* 19:1249–55.

Tsuruga, E., Takita, H., Itoh, H., Wakisaka, Y., and Kuboki, Y. 1997. Pore size of porous hydroxyapatite as the cell-substratum controls BMP-induced osteogenesis. *J Biochem* 121:317–24.

Wu, C., Fan, W., Zhou, Y., et al. 2012. 3D-Printing of highly uniform β-CaSiO3 ceramic scaffolds: Preparation, characterization and *in vivo* osteogenesis. *J Mater Chem* 22:12288–95.

Wu, C., Luo, Y., Cuniberti, G., Xiao, Y., and Gelinsky, M. 2011. Three-dimensional printing of hierarchical and tough mesoporous bioactive glass scaffolds with a controllable pore architecture, excellent mechanical strength and mineralization ability. *Acta Biomater* 7:2644–50.

Wu, C., Ramaswamy, Y., Boughton, P., and Zreiqat, H. 2008. Improvement of mechanical and biological properties of porous CaSiO3 scaffolds by poly(D, L-lactic acid) modification. *Acta Biomater* 4:343–53.

Wu, C., Zhang, Y., Zhu, Y., Friis, T., and Xiao, Y. 2010. Structure–property relationships of silk-modified mesoporous bioglass scaffolds. *Biomaterials* 31:3429–38.

Xia, W., and Chang, J. 2006. Well-ordered mesoporous bioactive glasses (MBG): A promising bioactive drug delivery system. *J Control Release* 110:522–30.

Xu, H. K., Weir, M. D., Burguera, E. F., and Fraser, A. M. 2006. Injectable and macroporous calcium phosphate cement scaffold. *Biomaterials* 27: 4279–87.

Xu, S., Li, K., Wang, Z., et al. 2008. Reconstruction of calvarial defect of rabbits using porous calcium silicate bioactive ceramics. *Biomaterials* 29:2588–96.

Yang, S., Leong, K., Du, Z., and Chua, C. 2002. The design of scaffolds for use in tissue engineering. Part II. Rapid prototyping techniques. *Tissue Eng* 8:1–11.

Yuan, H., Yang, Z., Li, Y., and Zhang, X. 1998. Osteoinduction by calcium phosphate biomaterials. *J Mater Sci: Mater Med* 9:723–6.

Zein, I., Hutmacher, D. W., Tan, K. C., and Teoh, S. H. 2002. Fused deposition modeling of novel scaffold architectures for tissue engineering applications. *Biomaterials* 23:1169–85.

Zhang, Y., and Zhang, M. 2002. Three-dimensional macroporous calcium phosphate bioceramics with nested chitosan sponges for load-bearing bone implants. *J Biomed Mater Res* 61:1–8.

5

Biomimetic Preparation and Biomineralization of Bioceramics

Zhongru Gou, Ahmed Ballo, and Wei Xia

CONTENTS

5.1 Introduction

Bioceramics are usually polycrystalline inorganic materials used in biomedical application, such as hard tissue repair and regeneration, drug delivery, and implant coatings. In this chapter, the concept of "bioceramic" is extended. Bioglass and glass-ceramic are also included as bioceramics. Table 5.1 summarizes various bioceramics that are acceptable and widely studied.

TABLE 5.1

Categories of Bioceramics

Category	Materials
Phosphate-based bioceramics	Hydroxyapatite (HAp)
	β-Tricalcium phosphate (β-TCP)
	α-Tricalcium phosphate (α-TCP)
	Octacalcium phosphate (OCP)
	Calcium-deficient hydroxyapatite (CDHA)
	Amorphous calcium phosphate (ACP)
	Calcium pyrophosphate (CPP)
Silicate-based bioceramics	Bioactive glasses
	Mesoporous bioactive glasses
	Calcium silicate
	Dicalcium silicate
	Tricalcium silicate
	Bioactive glass-ceramic
	Akermanite
	Hardystonite
Oxide bioceramics	Y_2O_3-ZrO_2
	Al_2O_3
	TiO_2
Nonoxide bioceramics	Si_3N_4
	SiC
	Carbon

Conventionally, the preparation of ceramics is a powder route followed by a sintering processing. In terms of glass and glass-ceramic, it is either a melting processing or sol-gel processing. Regarding bioceramics, because of "BIO" and super properties, scientists always try to learn from nature to prepare ceramics that have a similar structure and function of materials in biological systems, which can be named as "biomimetic preparation." Various biotemplates or structure-directing agents have been used and new techniques have been developed to fabricate ceramics that have (1) biological composition, (2) biomimetic structure, and (3) optimized mechanical strength. However, the mechanism behind the biomimetic preparation needs to be clarified. The following biomineralization is the process that modifies and functionalizes bioceramics by mimicking biologically controlled mineralization. The biological and clinical evaluations of these materials are also very important. It gives the feedback about the advantages and disadvantages of new bioceramics, and further directs how to design and better prepare materials.

5.2 Biomimetic Preparation of Bioceramics

In nature, there are numerous complex inorganic and inorganic–organic hybrid forms by natural evolution over millions of years. Calcium phosphates (CaPs) are the principal inorganic constituents of normal (bones, teeth, deer antlers, and some species of shells) and pathological (dental and urinary calculus and stones) calcifications. Nearly all hard tissues of the human body are made of CaPs. Structurally, they occur mainly in the form of poorly crystallized nonstoichiometric fluoride, sodium, magnesium, and carbonate-substituted hydroxyapatite (HAp) (Pasteris et al. 2008). Biomimetism of synthetic CaPs can be carried out at different levels, such as composition, structure, morphology, bulk, and surface chemical–physical properties. Biomaterials can copy these characteristics to not only optimize their interaction with biological tissues but also mimic biogenic materials in their functionalities. Therefore, mimicking "nature" and designing biomimetic bioceramics represent a promising way to reach technological innovations in biomedical fields, since biological materials exhibit a high degree of hierarchical architecture, hybridization, bioactivity, adaptability, and porosity. Detailed information on CaPs, their structure, chemistry, other properties, and biomedical application have been comprehensively reviewed (Dorozhkin and Epple 2002; Dorozhkin 2009). In this section some methods of synthesizing biomimetic CaP bioceramics will be described.

5.2.1 Transformation and Replication of Biological Origin into Calcium Phosphate Ceramics

Porous solids have the advantage of allowing circulation of body fluids and of increasing the potential for firm attachment of body tissue. Porous biomimetic CaP ceramics in simulating spongy bone morphology (porosity varying from a microporosity >1 μm to a macroporosity ranging from 100 to 1000 μm) has been prepared using various technologies to control pore dimension, shape, distribution, and interconnections. The earliest replamineform process (meaning replicated life forms) for porous metal and ceramic materials using natural marine invertebrates as templates was reported in the early 1970s (White et al. 1972). The porous calcium carbonate (CC) skeletal structure of some marine invertebrates has been achieved through specific building principles selected by evolution, and can be partially templated in man-made materials for the fabrication of porous bioceramic scaffolds for bone repair. The first reported coral-to-CaP conversion by employing exchange reactions at elevated temperatures and pressures was published by Roy et al. in the 1970s (Roy and Linnehan 1974). The concern of the work, of course, is the exchange process by which the porous coral structure preserved as HAp (containing some CO_3^{2-}) is formed. The coral-converted HAp constructs exhibited in the framework was around 130 μm in diameter, and their interconnections were 220 μm. The average porosity was 600 μm, and their interconnections were around 260 μm in diameter, which is favorable for the continued health of bony ingrowth (Shors 1999). The use of CaPs, since they are the most important inorganic constituents of hard tissues in vertebrates, is valid due to their chemical similarity to the mineral component of mammalian bone. It would, thus, have the advantage that it would be structurally, chemically, and mineralogically "identical" to normal human hard tissue, and stable in contact with tissue fluid. After that, some researchers took advantage of such hydrothermal reaction to prepare various carbon-containing HAp porous bioceramics. However, the partially converted coral-derived HAp/CC substrate with a "coating" of several micron depth of HAp on the porous resorbable framework of coral failed to induce the intrinsic induction of bone morphogenesis. Such partially converted HAp/CC biphasic biomimetic matrices did not induce bone formation even when reconstituted with naturally derived bone osteogenic proteins within a short time stage postoperatively. More recently, Ripamonti et al. (2009) implanted the HAp/CC constructs containing 5% and 13% HAp in heterotopic rectus abdominis sites. They found that induction of bone occurred in the concavities of the matrices at all-time points from 60 to 365 days. In particular, resorption of partially converted HAp/CC was apparent as well as remodeling of the newly formed bone. Northern blot analyses of samples from heterotopic specimens showed high levels of mRNA expression on BMP-7 and collagen type IV in all specimen types of 60 days, correlating with the induction of the osteoblastic phenotype in invading fibrovascular cells. This study demonstrates that partially converted HAp/CC constructs also induce spontaneous differentiation of bone, albeit only seen one year postimplantation (Ripamonti et al. 2009).

The echinoderm known as *Mellita eduardobarrosoi*, whose skeleton contains calcite as an inorganic constituent, can be used as the porous template for HAp bioceramics preparation as well. Araiza et al. (1999) investigated the influence of the initial pHs in the KH_2PO_4-KOH boiling solution on the formation of HAp during a nonhydrothermal treatment. Chemical and x-ray diffraction analysis of the resulting material showed that the conversion of calcite by this technique is mostly higher than 70%, yielding the highest conversion from pH 10 to 11. The obtained material is a mixture of the original calcite and HAp with varying stoichiometry and composition. Nevertheless, the interconnected porosity is preserved.

There are some other replication methods available to make porous CaP ceramics. The cancellous bone can be converted to HAp ceramic with high porosity after suffering high thermal treatment. The chemical composition, mechanical integrity, and macro- and microstructure of such HAp bioceramics (e.g., Endobon®) are similar to the natural bone mineral. In general, this kind of biomimetic HAp bioceramics possess a range of apparent densities from ~0.35 to ~1.44 g cm^{-3}, and the natural apatite precursor is not converted to pure HAp, but retains many of the ionic substituents found in bone mineral, notably carbonate, sodium, and magnesium ions. Such a material, porous, inorganic, and bioactive (containing some biologically essential trace minerals), should be highly compatible with body tissue, and when used as bone implants may become essentially integral with the bone. Moreover, the struts of the HAp material is not fully dense but have retained some traces of the network of osteocyte lacunae. Macrostructural analysis can demonstrate the complex interrelationship between the structural features of an open pore structure. It is also noted that the pore connectivity and mechanical strength are sensitive to macrostructural anisotropy and apparent density (Hing et al. 1999). Tancred et al. reported another negative–negative replica method that can also be used to fabricate the bonelike HAp porous bioceramics. Cancellous bone is used to create a negative replica; afterward acid is used to remove the bone and a HAp negative replica of wax mold is formed (Tancred et al. 1998).

5.2.2 Volatile Additives to Create Macropore Bioceramics

Macroporosity is usually formed due to the release of volatile materials and, therefore, the incorporation of pore creating additives is the most popular technique to create macropore bioceramics. The conventional porogens such as paraffin spheres, naphthalene, hydrogen peroxide, and polyvinyl butyral have been widely used. These additives are mixed to CaP powders or slurries. After molding, the organics burn away from the molding body during sintering. This approach allows direct control of the pore characteristics, which are a function of the amount and properties of the volatile phase. However, porous CaP ceramics processed by high-temperature treatment present a significant reduction of bioreactivity and growth kinetics of new bone due to the lack of resorbability. The low surface reactivity influence the osteogenic cell activation and osteoconduction (Dorozhkin 2010).

The conventional CaP bioceramics are strong but brittle, and thereafter are not suitable for most load-bearing applications. Bioceramics fail by sudden catastrophic fracture, which is not a trait that engineers find endearing. These materials behave in this way because they absorb very little energy during fracture. These difficulties could be alleviated if an energy-absorbing mechanism was built into the microstructure of materials to increase toughness. Previous reports of the extraordinary toughness and strength of mollusk shells have inspired much research because the shells, consisting of 99% CC, are hundreds of times tougher than simple polycrystalline limestone. The tough structure of shellfish nacre has been replicated in a ceramic by a simple water-freezing method (Deville, Saiz, Nalla, et al. 2006). Before that, the potential of freeze-casting as a means to create ceramics with a controlled and complex cellular architecture has started to attract research attention (Fukasawa et al. 2001), and the first reports on freeze-cast biomedical materials prepared from collagen solutions have appeared (Schoof et al. 1998, 2000, 2001). Thereafter, the freeze-casting of ceramic-based biomaterials has received a great deal of attention during the past few years. Current research focuses on both the fundamentals of the freeze-casting process, drawing on the well-established field of directional solidification of alloys, and the properties of the materials created (Deville 2008). This simple process, where a material suspension is simply frozen and then sublimated, provides materials with outstanding mechanical properties and unique porous architectures, where the porosity is almost a direct replica of the frozen solvent crystals. These remarkable properties are related to the fine-scale structure of the pore struts, a laminate of thin inorganic crystallite layers, and tough biopolymers, arranged in an energy-absorbing hierarchical microstructure, as recently reviewed by Deville (2010). It is thus a process with great promise for the development of biomimetic ceramics that emulate several design principles of the natural tissue that they are designed to replace.

Deville, Saiz, Nalla, et al. (2006) have shown how to replicate the nacre structure of shells using controlled freezing of mixtures of water and HAp ceramic powder. The composites made by this route have remarkably improved mechanical properties (e.g., up to 145 MPa for 47% of porosity and 65 MPa for 56% of porosity). The researchers achieve this by combining conventional manufacturing approaches as follows: (1) fine powder processing; (2) controlled solidification; and (3) freeze-drying. Their nacre-like composites start with a lamellar template assembled by ice crystals. Water freezes as lamellar dendrites, and the ice dendrites push the HAp particles into the interdendritic regions, making layers on the same scale as the ice. After the ice is removed by freeze-drying, the HAp ceramic keeps the shape of the interdendritic layers, forming a template for subsequent injection of tough polymer.

Fu et al. (2008b) and Rahaman and Fu (2008) investigated the relationship between HAp particle concentration, cold temperature, and solvent composition and porous microstructure. Freeze casting of aqueous suspension of HAp particles on a cold substrate produced porous constructs with a uniform, lamellar-type microstructure in which platelike HAp lamellas were oriented in the direction of freezing. Changes in the particle concentration of the suspension and temperature of the cold substrate did not drastically alter the lamellar-type microstructure but affected the porosity, the pore width, and thickness of the HAp lamellas. A decrease in the temperature of the cold substrate (–20°C to –196°C) caused a reduction in the size of the HAp lamellas and the pore cross-section. Addition of glycerol (~20 wt%) or dioxane (~60 wt%) to the aqueous solvent used in the suspensions produced finer pores and a larger number of dendritic structures, or much larger pores of cellular microstructure connecting the HAp lamellas. They also extended this study and investigated the effect of sintering conditions on the microstructure and mechanical behavior of freeze-cast HAp. Sintering (1250°C~1375°C) leads to a decrease in porosity (<5%) but an increase in strength of nearly 50%. The mechanical response shows high strain tolerance (5%–10% at the maximum stress), high strain to failure (>20%), and high strain rate sensitivity. The constructs with a porosity of 52% had compressive strengths of 12 MPa and 5 MPa in the directions parallel and perpendicular to the freezing direction, respectively. The favorable mechanical behavior of the porous constructs, coupled with the ability to modify their microstructure, indicates the potential of the present freeze-casting route for the production of porous scaffolds for bone tissue engineering (Fu et al. 2008b,c). Using the freeze-casting and drying technique, the biohybrid HAp/gelatine composites could also be prepared by infiltrating HAp lamellar scaffolds (45–55 vol.% of porosity) with a 10 wt% solution of gelatine. The HAp/gelatin porous lamellar scaffold showed appropriate compressive mechanical properties improved with the addition of gelatine: the strength increased up to 5 to 6 times, while the elastic modulus and strain approximately doubled (Landi, Valentini, et al. 2008).

Freeze-casting has quickly gained popularity as a manufacturing route for bioceramics because it is a comparatively straightforward physical process, based on biocompatible liquid carriers such as water. Next, the structure of freeze-cast materials, such as their overall porosity, pore size, and pore geometry, can easily be controlled across several length scales. Of equal importance is the ability to custom design both during and after freeze-casting the scaffold's cell-wall properties, such as surface roughness and chemistry, and with it the interface properties that are of critical importance to tissue–material interactions and a scaffold's successful tissue integration. These structural features, combined with the remarkable mechanical properties that directionally solidified materials offer despite their high overall porosity, are the basis of the great promise of freeze-cast biomaterials. Moreover, freeze-casting is ideally suited for the manufacture of bioceramics with property gradients. Because the porosity generated by freeze-casting is highly connected, it is possible to create hierarchical microstructures by sequential freeze-casting, thereby exploring the potential to introduce another level and another direction of porosity into the material. The high connectivity also makes it possible to coat the sample once or several times and to infiltrate it with another phase. Both offer great potential to integrate into the biomaterial growth factors to stimulate tissue ingrowth (Wegst et al. 2010).

Recently, wood-derived ceramic structures were obtained in a variety of compositions, through a sequence of chemical and thermal reactions. Pyrolysis is the starting process aimed at eliminating all the organic fraction of wood, leaving a porous skeleton in carbon, whose structure well reproduces the cellular organization of the native wood. Rattan wood in particular has a strong morphology similar to bone (Tampieri et al. 2001); it is characterized by a total porosity of 85% and large pores with diameter 250 ± 40 µm (Tampieri et al. 2009), evidencing a system of channel-like pores (simulating the Haversian system in bone) interconnected with a network of smaller channels (such as the Volkmann system). The criteria for the selection of rattan wood for bone scaffold development were based on the specifications of spongy bone.

5.2.3 Nanocrystalline Calcium Phosphate Bioceramics

The mineral phase of human bone and teeth consists primarily of substituted apatite crystals with size <100 nm. Synthetic HAp ceramic is chemically equivalent to these apatite crystals with subtle differences. Thus, HAp is a preferred material for bone repair because of its stability under *in vivo* conditions, compositional similarity, excellent biocompatibility, osteoconductivity, and ability to promote osteoblast functions. Nanocrystalline HAp is expected to have homogeneous resorption and better bioactivity than coarser crystals (Stupp and Braun 1997; Webster et al. 2001). Nanostructured biomaterials promote osteoblast adhesion and proliferation, osteointegration, and the deposition of calcium-containing minerals on the surface of these materials (Xu et al. 2005). Also, nanostructured bioceramics can be sintered at lower temperatures; thereby, problems associated with high temperature sintering, such as abnormal grain growth and formation of microcracks can be avoided. Properties and performance of HAp bioceramic can be improved by changing powder particle size and shape, their distribution, and agglomeration.

The biomimetism of bioceramics based on HAp could be carried out at different levels: composition, particle size and morphology, microstructure, and surface reactivity. The aim of a researcher is to realize a preparation of bioceramics that is biomimetic in all these characteristics in order not only to optimize the microstructure but also, and more ambitiously, to mimic the biogenic apatitic composition and particle dimensions in its functionality. This concept should be utilized in designing and preparing nanocrystalline biomimetic bioceramics in replacing hard tissues. In fact, the nanosize of biological tissue building blocks is one of the bases of their self-assemble ability and one that needs to be replaced by synthetic bioceramics in the synthesis of structured architectures on a multiple-length scale.

Many different methodologies have been proposed to prepare nanosized HAp (Schmidt 2000). They include wet chemical precipitation (Buehler et al. 2001), sol-gel synthesis (Kuriakose et al. 2004), solid state synthesis (Rao et al. 1997), coprecipitation (Kong et al. 2002), hydrothermal synthesis (Bonnelye et al. 2008), microwave processing (Siddharthan et al. 2006), high temperature flame spray pyrolysis, and several other methods by which nanocrystals of various shapes and sizes can be obtained. In general, the shape, size, and specific surface area of the HAp appear to be very sensitive to both the reaction temperature and the reactant addition rate. Depending upon the technique, materials with various morphology, stoichiometry, and level of crystallinity have been obtained. In order to optimize its specific biomedical applications, especially bioceramics function, the physical–chemical features that should be tailored in synthetic biomimetic HAp are dimensions, porosity, morphology, and surface properties.

Sintering of HAp is a complex process since many parameters influence the sintering process. Given its poor sinterability, HAp ceramics show low strength, especially in wet environments as would be found in physiological conditions. Obtaining full dense nanostructures is the key to enhance the mechanical and biological properties of HAp-based bioceramic materials. The mechanical properties and microstructures of the HAp bioceramics strictly depend on the characteristics of the original HAp powder, including crystallinity, agglomeration, stoichiometry, and substitutions, and on the processing conditions. Nanostructure processing can improve the sinterability of bioceramics and enhances the mechanical reliability by reducing flow sizes. The high volume fraction of grain boundaries in nanocrystalline bioceramic compacts also provides for increased ductility and superplasticity for low-temperature net-shape forming. Recent studies have reported that the nanocrystalline HAp materials produced were thermally stable up to 1300°C, whereas conventional commercial HAp powders could not be sintered to >70% theoretical density via pressureless sintering and underwent decomposition to α-tricalcium phosphate by 900°C. Fully dense HApP compacts can be achieved at temperatures as low as 900°C, while retaining the ultrafine microstructure (Ahn et al. 2001).

Nanostructured bioceramics are usually processed by compacting nanopowders at high pressures, and sintering at different times and temperatures and in various atmospheres. Pressure-assisted methods, such as hot pressing, hot isostatic pressing, and sinter forging, are also applied to obtain nanostructured ceramic materials (Groza 1999; Raynaud et al. 2002; Veljović et al. 2009). Hot pressing makes it possible to enhance densification kinetics and limit grain growth. This technique is usually used to process HAp ceramics with a controlled microstructure. After the sintering and hot pressing, the calcium deficient apatite turns into a mixture of HAp and β-tricalcium phosphate (β-TCP). A high amount of β-TCP is highly detrimental to the sintering and mechanical properties of HAp bioceramics (Raynaud, Champion, Bernache-Assollant, et al. 2002). In order to avoid degradation of the mechanical properties, the absence of the β-TCP in HAp bioceramics is advisable.

5.2.4 Trace Element Incorporating into Calcium Phosphate Bioceramics

One of the major drawbacks of stoichiometric CaPs is their inferior osteogenic capacity and poor mechanical strength compared with the living bone tissue, and this has been attributed to the subtle but significant chemical difference found in the structure. Natural bone apatite is nonstoichiometric and contains relatively higher levels of magnesium, sodium, and carbon (in the form of carbonate groups, CO_3^{2-}), and lower levels of trace minerals such as Mg, Sr, Zn, and F. These trace ions at critical levels are considered to play pivotal roles in the process of biomineralization as well as other diverse effects on nanocrystal size, dissolubility, and bioactivity of synthetic bioceramics. In recent years, a large number of studies have been reported on the synthesis of biologically essential trace mineral-doped CaPs (Boanini et al. 2010). Bodhak et al. (2011) reported that the combined addition of MgO and SrO in commercially procured phase pure HAp ceramics was found to be the most beneficial in enhancing the polarizability of pure HAp by inhibiting high-temperature HAp phase decomposition, and established the significance of dopants on polarized HAp for bone graft applications. Sol-gel chemistry can be used to prepare Sr-doped CaP ceramics exhibiting a porous structure. Doping with strontium ions has a clear effect on the proportions of the different CaP phases, increasing the amount of β-tricalcium phosphate (β-TCP). Strontium ions also substitute for calcium in both HAp and TCP in specific sites (Renaudin et al. 2008). Meanwhile, latteric parameters (a, c) unit cell volume and density in Sr-substituted HAp (Sr-HAp), $Sr_xCa_{1-x}(PO_4)_3OH$ are shown to increase linearly with strontium addition and are consistent to a maximum at $x = 0.5$, then decrease with the increase of Sr content (O'Donnell et al. 2008). Therefore, the solubility of Sr-HAp increases with increasing Sr content and this can be interpreted as a destabilization of the crystal structure by the larger strontium ion (Pan et al. 2009).

Zinc can also be doped into the CaP structure by a wet chemical method in aqueous solutions. The apatite lattice parameters and phase changes with the inclusion of zinc due to the partially substituted Ca ions in the apatite structure (Ren et al. 2009). Variation in the amount of ZnO present in TCP and HAp ceramics shows differences in sintering behavior. At a sintering temperature of 1250°C both TCP and HAp experience an increase in densification with increasing ZnO content. The ceramics sintered at 1250°C proved to be harder than those sintered at 1300°C. Analysis of microstructure using SEM can reveal that grain size changes due to ZnO doping in TCP but have limited influence on HAp (Bandyopadhyay et al. 2007).

Silicon (Si) substitution in the crystal structures of calcium phosphate (CaP) ceramics generates materials with superior biological performance to stoichiometric counterparts (Vallet-Regi and Arcos 2005). Si, an essential trace element required for healthy bone and connective tissues, influences the biological activity of CaP materials by modifying material properties and by direct effects on the physiological processes in skeletal tissue (Pietak et al. 2007). The synthesis of Si-HAp and Si-α-TCP has focused on wet chemical methods where Si is introduced as a chemical carrier such as tetrapropyl orthosilicate (TPOS) or tetraethyl orthosilicate (TEOS), Si IV acetate (Si(COOCH3)4) (Gibson et al. 1999, 2002; Kim et al. 2003), or as some form of nanoparticulate silica during the precipitation or firing of an amorphous CaP or nanocrystalline HAp (Reid et al. 2005; Li et al. 2006). Hydrothermal (Tang et al. 2005) and solid state (Boyer et al. 1997) methods of preparation have also been investigated. The materials are typically sintered at temperatures between 700°C and 1200°C. TCP, unlike HAp, do not precipitate from solution but rather are created from decomposition reactions at temperatures exceeding 700°C. A calcium-deficient HAp or amorphous CaP material with a Ca/P ratio between 1.5 and 1.67 will decompose into β-TCP or a biphasic system of β-TCP and HAp with sintering between 700°C and 1125°C. Sintering above 1125°C, the α-polymorph of TCP, is the stable phase. However, the presence of Si stabilizes the α-TCP polymorph, which can form at lower sintering temperatures of 700°C (Langstaff et al. 1999; Sayer et al. 2003; Reid et al. 2005).

5.2.5 Inorganic–Organic Composite Biomimetic Bioceramics

CaP bioceramics have already been used in clinics for filling of bone defects due to their bone-bonding ability with living tissue. However, the brittleness and low fatigue strength in a physiological environment limit their use for load-bearing repair or substitute. In recent years, composite materials comprised of bioactive inorganic CaP particles and organic (bio)polymers have been studied, such as HAp/Col, HAp/PLA, and HAp/gelatin. Narbat et al. (2006) studied the fabrication of porous hydroxyapatite-gelatin composite scaffolds for bone tissue engineering. It was observed that the prepared scaffold has an open, interconnected porous structure with a pore size of 80 to 400 μm, which is suitable for osteoblast cell proliferation. It was found that the GEL/HAp with ratio of 50 wt% HAp has the compressive modulus of ~10 GPa, the ultimate compressive stress of ~32 MPa, and the elongation of ~3 MPa similar to that of trabecular bone. Wang et al. (2002) reported that the nanoscale HAp (n-HAp)/PAA_{66} composites were similar to bone apatite in size, phase composition, and crystal structure. The biomimetic n-HAp crystals were uniformly distributed in the polymer matrix and its content can reach 65%, close to that in natural bone. Bending strength, tensile strength, and impact strength of n-HAp/PA66 composite with a n-HAp content of 55 wt% were higher than those of μ-HAp/PA66 with the same HAp content, increased by 31.3%, 38.9%, and 68.0%, respectively. The bending modulus of n-HAp/PA66 composite is 6.2 GPa, similar to the range of natural bone (from 6.9 to 27.4 GPa).

In addition, the method to coat the porous CaP scaffold with a polymer lining is appropriate to improve the mechanical properties of the interconnected porous CaP scaffold, while the high porosity and interconnectivity is maintained. Usually, synthetic polymers are preferable lining candidates to a natural-based polymer for their predictable and reproducible degradation characteristics as well as reasonable mechanical properties. For example, polycaprolactone (PCL) is a synthetic biocompatible polymer with higher fracture energy than other biopolymers. Thus, the porous HAp scaffold coated with a PCL lining and biomimetic CaP to form a composite scaffold can possess an improved toughening properties (Zhao et al. 2008).

5.3 Biomineralization of Bioceramics

Biomineralization, or biomimetic mineralization, refers to the process by which living forms precipitate mineral materials. The common idea is that biomineralization is the biologically controlled mineralization. However, the concept has been "misused" or widened. Any mineralization process that forms minerals with biological morphology or biological effect has been called "biomineralization," even if there is no "bio-stuff" involved in the whole process. In this chapter, the concept of biomineralization covers both real biologically controlled and nonbiological, or self-assembled, mineralization.

Biomineralization of bioceramics can be considered to form bonelike apatite on their surfaces via a biologically or nonbiologically controlled precipitation process. Hench (1991) first reported the formation of an apatite layer on the bioglass surface. It is not just a biomineralization process, but bonds between bone and bioactive materials and avoids soft tissue surrounding an artificial biomaterial.

There are two possible ways of crystallizations via a biomineralization process (Colfen and Mann 2003). One is thermodynamic control and another is kinetic control (see Figure 5.1). Thermodynamic crystallization is a one-step process. The fast growing faces have high surface energies and they will vanish in the final morphology and vice versa. The kinetic crystallization is based on the modification of the activation-energy barriers of nucleation, growth, and phase transformation. It often involves an initial amorphous

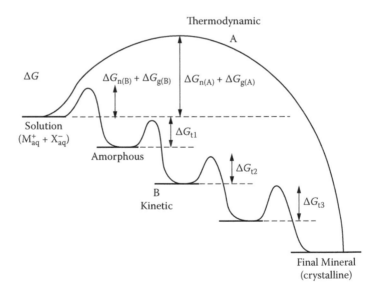

FIGURE 5.1
Crystallization pathways under thermodynamic and kinetic control. Whether a system follows a one-step route to the final mineral phase (pathway A) or proceeds by sequential precipitation (pathway B), depends on the free energy of activation (DG) associated with nucleation (n), growth (g), and phase transformation (t). Amorphous phases are common under kinetic conditions. (From H. Colfen and S. Mann, *Angewandte Chemie-International Edition* 42(21): 2350–2365, 2003.)

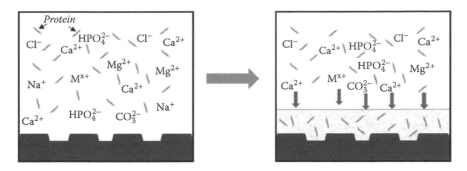

FIGURE 5.2
(See color insert.) Schematic process of biomineralization of biological hydroxyapatite in simulated body fluid containing proteins.

phase that may be nonstoichiometric, hydrated, and susceptible to rapid phase transformation (Colfen and Mann 2003).

Generally, the biomineralization of bioceramics is always under mild conditions (see Figure 5.2). The temperature is around 37°C, close to the human body temperature. The compositions of solutions are similar to that of human body fluid. Proteins are involved or not involved in the crystallization. The formation consists of apatite with small crystals and low crystallinity. The biomineralized apatite does not contain calcium and phosphate only. It is a multi-ion substituted apatite, and can be easily functionalized by proteins and drugs.

Table 5.2 summarizes most of the popular mineralizing solutions that mimic the compositions of human body fluid. Simulated body fluid (SBF) plays a significant role in biomineralization of bioceramics, which is affected by the composition, concentration, pH, and flowing state of SBF and its additives, such as trace elements, proteins, and drugs. The most famous SBF is Kokubo's solution (Kokubo et al. 1990), which has almost the same mineral concentrations as human blood plasma and is used to evaluate the ability of the bone bonding in an *in vitro* model. Other modified Kokubo's SBF have also been used to study the apatite formation on a bioceramic surface. Except for Kokubo's SBF, other buffer solutions containing calcium and phosphate ions have been tested, such as ringer's solution, EBSS (Earle's balanced salt solution), Hank's solution, SCS (supersaturated calcification solution), and DPBS (Dulbecco's phosphate buffer saline).

Biomineralization of bioceramics is also interesting. Because it mimics the process of bone mineralization, the newly formed crystal is similar to the apatite in human bone. Thus, the bone-like apatite precipitated from the above solutions is expected to show high affinity to the bone tissue. The mechanism behind the biomineralization of bioceramics would be useful for the explanation of (1) bone mineralization, (2) bonding between bond and bioactive ceramics, and (3) how to decrease the failure of bone materials.

TABLE 5.2

Ion Concentrations of Simulated Body Fluids and Human Blood Plasma

	Na^+	K^+	Mg^{2+}	Ca^{2+}	Cl^-	HCO_3^-	HPO_4^-	SO_4^{2-}
Human blood plasma (Gamble 1967)	142.0	5.0	1.5	2.5	103.0	27.0	1.0	0.5
Ringer (Ringer 1883)	130	4.0		1.4	109.0			
EBSS (Earle 1943)	143.5	5.37	0.8	1.8	123.5	26.2	1.0	0.8
HBSS (Hanks and Wallace 1949; Hanks 1975)	142.1	5.33	0.9	1.26	146.8	4.2	0.78	0.41
SCS (Wen et al. 1997; Habibovic et al. 2005)	140.4, 136.8	3.71		3.1	142.9, 144.5		1.86	
Dulbecco's PBS	136.9	2.68	0.49	0.9	142.4		9.57	
SBF (Kokubo et al. 1990)	142.0	5.0	1.5	2.5	148.8	4.2	1.0	0.5
mSBF (Kokubo 1991; Ohtsuki et al. 1991; Cho et al. 1995; Bigi et al. 2000; Tas 2000; Oyane et al. 2003)	142.0	5.0	1.0, 1.5	1.6, 2.5	147.8, 125.0, 124.5, 103.0	10.0, 27.0, 4.2	1.0	0.5
SBFx5 (Barrere et al. 2000; Habibovic et al. 2002)	714.8, 704.2		7.5, 1.5	12.5	723.8, 711.8	21.0, 10.5	5.0	

5.3.1 Biomineralization of Oxide Bioceramics

The surface functional groups, such as Si-OH, Ti-OH, Zr-OH, Ta-OH, and Nb-OH, could induce the growth of calcium phosphate in a body environment (Lee et al. 2006). The surface potential of these functional groups is initially negative owing to their isoelectric point in blood plasma (pH 7.4). The nucleation is assumed to involve electrostatic interaction of the surface functional groups with Ca^{2+} and phosphate ions.

Preparation methods, crystallinity, and surface treatment will influence the formation of these active groups in SBF. Sol-gel–derived silica can induce the bonelike apatite precipitation on its surface, but silica glass and quartz do not (Li et al. 1994). Hydrated silica in the fluid plays the key role in the nucleation of apatite.

Sol-gel–derived TiO_2 can also induce the apatite formation because of the formation of Ti-OH groups in SBF (Li et al. 1994; Wei et al. 2002). The ability of apatite formation on titania surfaces with different crystallinity is different (Uchida et al. 2001). Sol-gel–derived anatase is more active than sol-gel–derived rutile and amorphous titania. Regarding alumina, although sufficient Al-OH groups may remain in the alumina gel, they do not initiate apatite generation when immersed in SBF (Li et al. 1994). That means not all Ti-OH groups, but certain types of Ti-OH groups in a specific structural arrangement, are effective in inducing apatite nucleation. The crucial part is how well the interface between the organized hydroxylated surface and HAp nuclei structurally match, and also the surface charge (Uchida et al. 2001). Hydroxylated 110 (anatase) and 0001 (HAp) match their interface via three parts: (1) hydrogen bond interaction, (2) crystal lattice matching, and (3) stereochemical matching. Lindberg et al. (2008) reported experimental observations of early growth and growth of apatite on single-crystal rutile substrates (100), (001), and (110). The adsorption of calcium and phosphate ions was faster on the (001) and (100) surfaces than on the (110) surface in the early stage (Lindahl et al. 2010). After a long time soaking, the hydroxy-apatite precipitate nucleus on the (001) surface led to faster coverage of this surface compared to the (110) and (100) rutile surfaces. The ion-substituted apatite can form on pretreated rutile surfaces. Xia, Lindahl, Persson, et al. (2010) reported that the ion doping not only changed the composition but also influenced the morphology. Strontium-, silicon-, and fluoride-substituted apatite appeared spherical, nano-flake, and needle-like, respectively.

5.3.2 Biomineralization of Silicate- and Aluminate-Based Bioceramics

It is known that simulated body fluid contains calcium and phosphate ions that are supersaturated with respect to apatite. However, apatite could not spontaneously precipitate under a normal condition. As described in Figure 5.1, the energy barrier for the apatite nucleation inhibits the initial formation. Hench (1991) have reported that the formation of Si-OH group on bioactive glass surfaces plays the key role in the nucleation of apatite. The formation of Si-OH group is triggered by the release of calcium and sodium ions from bioactive glasses. The release of cations also increases the degree of supersaturation of SBF. Both the formation of Si-OH and the release of cations help the nucleation and growth of apatite. Figure 5.3 illustrates the nucleation and growth of bonelike apatite (Lee et al. 2006). Normally other ions, such as carbonate and Mg, will spontaneously precipitate with calcium and phosphate.

Except for bioactive glasses, in terms of silicate-based bioceramics, mesoporous bioactive glass (MBG) is another interesting material. MBGs constitute a new family of bioceramics with the fastest *in vitro* bioactivity studied so far. The interest of MBG is due to their specific structure with high surface area and suitable pore size and pore volume. The extreme high surface area can offer more places for the formation of Si-OH group that help the nucleation of apatite. Those result in a fast precipitation of apatite on MBG surface. A sequential transition from amorphous calcium phosphate (ACP) to octacalcium phosphate (OCP) and to calcium deficient carbonate hydroxyapatite (CDHA) maturation, similar to the *in vivo* bone biomineralization, has been observed in a MBG system with 2D hexagonal structure and high Ca content (37 mol%) (Izquierdo-Barba, Arcos, et al. 2008). It was the first time to find the ACP-OCP-CDHA maturation sequence in a bioactive ceramic system. The intense exchange of Ca^{2+} and H_3O^+ decreases the local pH value that favors the formation of OCP.

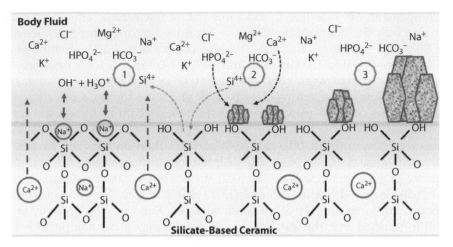

1. Rapid diffusion-controlled ion exchange of network modifying Ca^{2+} and/or Na^+ ions from the ceramic with H_3O^+ ions from the body fluid and silanol groups (Si-OH) at the surface form.
2. HPO_4^{2-} and Ca^{2+} ions from the body fluid are incorporated into the silicate layer from an amorphous calcium phosphate layer (nucleation).
3. Crystallization and growth into biological hydroxyapatite.

FIGURE 5.3
(See color insert.) Schematic of the mechanism of formation of calcium phosphate on the surface of silicate based ceramics.

Arcos et al. (2011) also reported that the biomineralization of MBGs could enhance their mechanical strength. The increase of the mechanical strength is due to their internal mesoporous structure, which facilitates the ionic transport through the bulk pieces. The newly crystallized apatite phase reinforces the grain boundary interaction, resulting in a biomimetic nanocomposite within the whole implant volume.

The mechanism of precipitation of apatite on the family of calcium silicates ceramics is similar to silicate-based bioactive glasses. The release of high content of Ca ion and the formation of Si-OH on their surfaces stimulate the nucleation of apatite crystals. Chang and colleagues have proved the good *in vitro* bioactivity of calcium silicate, dicalcium silicate, tricalcium silicate, and other calcium silicate-based ceramics (Gou and Chang 2004; Gou et al. 2004, 2005; Wu and Chang 2004; Zhao and Chang 2004, 2005; Lin et al. 2005). Li et al. (2007) reported that mesoporous amorphous calcium silicate had a fast precipitation of apatite within 4 hours on its surface.

Ca-aluminate (CA)-based bioceramics have been studied as a biomaterial in odontology, orthopedics, and as a carrier material for drug delivery (Hermansson 2011). CA-based materials exhibit unique curing and hardening characteristics. The hydrate layer forms when CA meets with the fluid. After a long time soaking, the katoite would transform into apatite and gibbsite phases (Hermansson et al. 2008).

FIGURE 5.4
SEM images of SBF-soaked (a) β-TCP and (b) HA.

5.3.3 Biomineralization of Phosphate-Based Bioceramics

Phosphate-based bioceramics, such as the calcium phosphate family, are one of the famous materials in biomedical application. Typically, calcium and phosphate ions are always involved in a bone mineralization. Supersaturated calcium and phosphate ions are induced by collagen to form from ACP to OCP to CDHA. The calcium phosphate family itself is biomineral. Furthermore, they can still induce apatite precipitated on their surfaces. Compared to the silicate-based bioceramics, the precipitation rate is slow. It is a dissolution and reprecipitation process. The release of calcium and phosphate ions from the surfaces will determine the mineralization rate. That means not only the phase composition but also surface area and preparation methods will influence the deposition. The final deposition is not the original phase, but a bonelike apatite that is calcium deficient, multi-ion substituted, and nano-sized (see Figure 5.4). Therefore, a completely crystalized and stoichiometric hydroxyapatite is not a good material that can induce new mineral depositing on its surface if there is not any surface functionalization. A-W glass-ceramic is a biphase ceramic containing apatite and calcium silicate. The mechanism of apatite deposition on A-W surface is similar to that on bioactive glasses (Kokubo et al. 1992). However, the silica gel layer has not been observed on its surface after apatite forms because there is not a continuous calcium silicate phase along the surface (Ohtsuki et al. 1995).

5.4 *In Vitro* and *In Vivo* Reaction of Biomimetic and Biomineralized Bioceramics

5.4.1 Clinical Background

For many years, the clinical application of bioceramics in medicine and dentistry has been largely limited to nonload bearing parts of the skeleton due to their poor mechanical properties. Bioceramics have been used as an alternative to autografts and allografts. They include bioactive glass, calcium carbonate, calcium sulfate, and calcium phosphates of biologic (derived from bovine bone, coral, and marine algae) or synthetic origin. These bioceramics are available as granules or blocks (dense or porous), with various designed shapes and sizes, cements, or as coatings on titanium implants. In the case of load-bearing applications, bioceramic coatings are applied onto a metal substrate in order to combine the strength of metal with the bioactivity of bioceramics.

Based on the nature of the attachment of the bioceramics to the living bone tissue, bioceramics are described as either bioinert (makes contact with bone without leading to low tissue reactions and formation of a fibrous layer and with no chemical contacts between bone and materials) or bioactive (directly attached to the bone by chemical strong bonding without interposition of a fibrous layer) (Cao and Hench 1996). Commercial alumina (Al_2O_3) and zirconia (Zr_2O_3) that are used for both dental and orthopedic applications are considered as stable bioinert bioceramics. In contrast, bioactive bioceramic can be biostable (i.e., calcium phosphate) or bioresorbable (i.e., bioactive glasses and glass-ceramics).

Bone is a dynamic, vascular, living tissue that is simply described as a biocomposite consisting of organic and inorganic parts. The inorganic is mainly crystalline mineral salts and calcium, which is present in the form of hydroxyapatite, $Ca_{10}(PO_4)_2(OH)_2$ (Beevers and McIntyre 1946). HAp in bone is a multisubstituted calcium phosphate, including traces of CO_3^{2-}, F^-, Mg^{2+}, Sr^{2+}, Si^{4+}, and so on (Beevers and McIntyre 1946; Dorozhkin and Epple 2002; Vallet-Regi and Gonzalez-Calbet 2004).

Critical physic-biochemical properties of bone include (1) interconnecting porosities (macro- and microporosity), (2) biodegradability (bone remodeling), (3) bioactivity, (4) osteoconductivity, and (5) osteoinductivity. These bone properties have emulated in the bioceramics synthesis. For example, osteoinductivity is introduced by mixing the bioceramics with osteogenic molecules (e.g., growth factors, demineralized bone matrix). However, differences in composition and syntheses or processing methods affect the properties of the bioceramics.

Among the various techniques that have been used to coat Ti implants with layers of hydroxyapatite (de Groot et al. 1987), calcium phosphate (de Groot 1989), or mixtures of the two (Klein et al. 1994), the most common is the plasma-spray technique. These coated implants are, moreover, characterized by a rough surface profile, which further improves osteoconduction and osseointegration. However, plasma-sprayed HAp coatings are approximately 30 to 50 mm in thickness and are prone to delaminate from the metal substrate in certain situations owing to their poor bonding strength, which creates a weak interface that may eventually lead to implant failure (Lemons et al. 1988; Albrektsson et al. 1992; Gotfredsen et al. 1995).

To avoid the drawbacks of plasma-sprayed HAp coatings, scientists have developed a new coating method inspired by the natural process of biomineralization. The biomimetic deposition of calcium phosphate onto surfaces of implant materials is the most successful example of biological mimicry, or biomimetics, in fields of osseointegration. This technique tried to mimic some of the properties of bone by adjusting the composition, introducing interconnecting porosity, and incorporating osteoinductive factors or biologically active molecules. This technique was originally developed by Kokubo et al. (1990). This method allows HAp and other calcium phosphate surfaces to be deposited on substrates in a SBF under physiological conditions of temperature and pH (Kuroda et al. 2004), on complex geometrical shapes, primarily to improve their biocompatibility and biodegradability.

5.4.2 Biological Performance of Biomimetic and Biomineralized Bioceramics

Regardless of the nature of the biomaterials, *multiple events occur immediately after* osteotomy and implant installation. The peri-implant wound healing process starts with extravasation (bleeding), so that blood is the first tissue that comes in contact with the implant surface after implantation. This contact results in a series of immediate and early processes including adsorption of proteins and other molecules from the biological microenvironment onto its surface (Shard and Tomlins 2006; Thevenot et al. 2008). The next stage involves the interaction of cells with the "surface" of the implant via the adsorbed protein layer. The cell-protein bound surface interface, occurring from as short as minutes after and up to days following implant placement, initiates cellular adhesion, migration, and differentiation, which occurs from a few hours to several days after implantation (Wilson et al. 2005). Erythrocytes, thrombocytes (platelets), and leukocytes are the first cellular participants at the interface and, consequently, a dense fibrin network is formed, resulting in provisional matrix formation. This stage is tightly regulated by numerous biological factors, including extracellular matrix proteins, cell surface-bound and cytoskeletal proteins, by chemical characteristics and topographies at the implant surface, and by the released ions and products from the material (Ratner and Bryant 2004). Further, inflammatory cells such as monocytes or macrophages can be seen at a very early stage at the bone interface with HAp surfaces (Muller-Mai et al. 1990) and they have been regarded as a pivotal cell lineage both during successful osseointegration and in the pathogenesis of implant failure (Aubin 1998). Thereafter, at the end of the first week of implantation, callus and mesenchymal tissues will have entirely replaced blood while host bone resorption has started. Finally, between the second and fourth weeks of implant installation, callus, mesenchymal tissues, and host bone will have gradually disappeared in favor of newly formed bone while bone remodeling takes place.

The biomaterials features can affect the molecular and cellular interactions at their surface and consequently can affect the process of bone formation. Calcium phosphate bioceramics are integrated within bone by a unique phenomenon that involves complex interactions between multiple cellular and molecular events that regulate the healing process at the interface (Frayssinet et al. 1993). Calcium phosphate bioceramics have such compositional resemblance to bone mineral that they induce a biological response similar to the one generated during bone remodeling. Calcium phosphate bioceramics are considered to be bioactive and osteoconductive. Bioactivity would be due to epitaxial nucleation of carbonated apatite crystals at the surface of ceramic grains. This layer of biological apatite might contain endogenous proteins and might serve as a matrix for osteoprogenitors cell attachment and growth (Davies 2003).

Biomimetic precipitation occurs on a variety of bioceramics such as bioglass (Leonor et al. 2002), bioactive fiber-reinforced composite implant (Ballo et al. 2009), sol-gel processed silica, and titania (Uchida et al. 2001). The ability of a bioceramic material to induce a biomimetic precipitation has been correlated to the degree of bioceramics dissolution and ions released at the interface (Ducheyne et al. 1993). This dissolution process is highly dependent on the nature of the bioceramic phosphate materials (composition, porosity, surface area, and crystallinity) (LeGeros 1993; Christoffersen et al. 1997; Barrere et al. 2003) and on the composition and supersaturation of the environment *in vitro* (Hyakuna et al. 1990), or the implantation site (Daculsi et al. 1990; Barralet et al. 2000). The morphologies of the surface prepared by the biomimetic method in a SBF are different from the normal coating derived from the sol-gel and plasma-spray process. Instead of being dense and smooth, they are rough and porous and thus have morphology of platelike particles that may be favorable for the adhesion and proliferation of cells (Brunette and Chehroudi 1999; Ito 1999).

Biomimetic calcium phosphate materials have previously been evaluated both *in vivo* and controllable *in vitro* experiments. These materials promote surface adhesion and proliferation, which contributes significantly to good osteoblastic cell biocompatibility (Vaahtio et al. 2006; Zhang et al. 2009). It has also been reported that biomimetic calcium phosphate coatings are more soluble in physiological fluids and more resorbable by osteoclasts than high-temperature plasma-sprayed hydroxyapatite coatings (Leeuwenburgh et al. 2001; Barrere et al. 2003; Wijenayaka et al. 2009). Thus, these materials might be useful to enhance favorable bone remodeling, an important process in bone healing involving osteoclastic resorption and subsequent bone formation by osteoblasts.

The osseointegration of titanium implants coated with biomimetic calcium phosphate has been investigated in experimental animal models. These studies have demonstrated a greater bone-implant contact for biomimetic calcium phosphate coatings than for uncoated titanium implants (Barrere et al. 2003; Habibovic et al. 2005). Compared with other surface treatments, the biomimetic calcium phosphate coating appears to work as an accelerator of bone ingrowth compared to a conventional titanium plasma spray coating in a preclinical model (Biemond et al. 2011). Interestingly, it has been demonstrated that the physiological ions substitution in the crystal structures of CaP ceramics such as HAp and TCP generates materials with surface bioactivity and superior biological performance comparing traditional hydroxyapatite and tricalcium phosphate materials (Landi et al. 2008; Xia, Lindahl, Persson, et al. 2010; Bose et al. 2011; Ballo et al. 2012). Some of these ions can affect the crystal lattice, and therefore can accelerate the dissolution, for example, carbonate, silicate, or strontium in HAp. On the other hand, some additives reduce *in vitro* and *in vivo* the dissolution process, for example, fluoride in HAp, magnesium, or zinc in β-TCP (Okazaki et al. 1982; Dhert et al. 1993; Porter et al. 2004).

Recently, our finding has suggested that the incorporation of silicon (Si) and strontium (Sr) ions in the HAp biomimetic (Si-HAp and Sr-HAp, respectively) surfaces improved the surface bioactivity and stimulated bone apposition in the very early stages of bone healing following implant placement, leading to enhanced osseointegration along the surface of implants (Figure 5.5) (Ballo et al. 2012). A possible explanation of the higher bioactivity of Si-HAp and Sr-HAp surfaces is that Si-HAp ceramic has a higher dissolu-

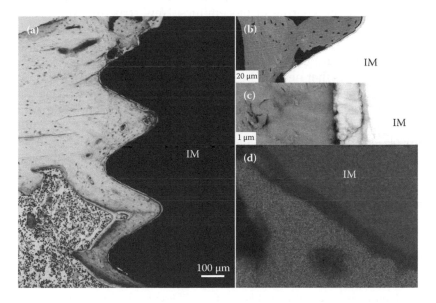

FIGURE 5.5
(See color insert.) Sr-HA implants after 28 days of implantation in animal model. (a) Histological sections revealed mineralized bone growth in the medullary area along and in direct contact with surfaces was observed (contact osteogenesis). (b, c) Backscatter scanning electron microscopy micrographs of the implants after 28 days of healing. (b) Low-magnification image showing the implant, implant surface, and bone tissue. Osteocyte lacunae and canaliculi were frequently observed close to the implant surface. (c) Higher-magnification image showing direct contact was observed. (d) Overlapped element maps of calcium (green), titanium (red), and oxygen (blue), showing bone formation along the HAp-implant surface at 28 days healing. The enlarged surface oxide is shown in purple (overlapped blue and red) along the implant perimeter.

tion rate than HAp ceramic owing to the low connectivity of Si incorporated in HAp, which facilitates the release of Si into the contact fluid.

Furthermore, it has been observed that the dissolution rate of the surface coating is critical to the cell attachment and proliferation (Muller-Mai et al. 1990). It is suggested that the cell attachment and cell growth could be improved by adjusting the balance between the crystallinity and the biosolubility of the Si-HA coating (Zhang et al. 2009). The synergic effect, by adding the Sr ion into the crystal structure of CaP, has been explained by the concept of Sr possibly increasing bone formation and reducing bone loss. Sr is known to reduce the proliferation and differentiation of osteoclasts, which generally reduces resorption of bone (Buehler et al. 2001; Bonnelye et al. 2008).

5.5 Conclusion and Comments

Biomimetic preparation and biomineralization are promising approaches of the design and engineering of bioceramics. New structure and properties of bioceramics will be figured out even for "old materials," such as calcium phosphates, calcium silicates, and bioactive glasses. Even though we already see the bright future of this "BIO-technique," as we understand, there are still some aspects that need to be considered: (1) biomimetic effects of new bioceramics, such as mechanical properties and material and tissue interaction; (2) the mechanism of biomimetic preparation and biomineralization at the nano and atomic level; (3) proper set up of *in vitro* and *in vivo* evaluation protocols and performance of biological evaluations.

References

E. S. Ahn, N. J. Gleason, A. Nakahira, et al. 2001. Nanostructure processing of hydroxyapatite-based bioceramics. *Nano Letters* 1(3): 149–153.

T. Albrektsson, P. Astrand, W. Becker, et al. 1992. Histologic studies of failed dental implants: A retrieval analysis of four different oral implant designs. *Clinical Materials* 10: 225–232.

M. A. Araiza, J. Gómez-Morales, R. R. Clemente, et al. 1999. Conversion of the echinoderm *Mellita eduardobarrosoi* calcite skeleton into porous hydroxyapatite by treatment with phosphated boiling solutions. *Journal of Materials Synthesis and Processing* 7(4): 211–219.

D. Arcos, M. Vila, A. Lopez-Noriega, et al. 2011. Mesoporous bioactive glasses: Mechanical reinforcement by means of a biomimetic process. *Acta Biomaterialia* 7(7): 2952–2959.

J. E. Aubin. 1998. Bone stem cells. *Journal of Cellular Biochemistry (Supplement)* 30–31: 73–82.

A. M. Ballo, E. A. Akca, T. Ozen, et al. 2009. Bone tissue responses to glass fiber-reinforced composite implants—A histomorphometric study. *Clinical Oral Implants Research* 20(6): 608–615.

A. M. Ballo, W. Xia, A. Palmquist, et al. 2012. Bone tissue reactions to biomimetic ion-substituted apatite surfaces on titanium implants. *Journal of the Royal Society, Interface* 9(72): 1615–1624.

A. Bandyopadhyay, E. A. Withey, J. Moore, et al. 2007. Influence of ZnO doping in calcium phosphate ceramics. *Materials Science and Engineering* C 27(1): 14–17.

J. Barralet, M. Akao, H. Aoki, et al. 2000. Dissolution of dense carbonate apatite subcutaneously implanted in Wistar rats. *Journal of Biomedical Materials Research* 49(2): 176–182.

F. Barrere, P. Layrolle, C. A. van Blitterswijk, et al. 2000. Fast formation of biomimetic Ca-P coatings on Ti6Al4V. *Mineralization in Natural and Synthetic Biomaterials* 599: 135–140.

F. Barrere, C. M. van der Valk, R. A. Dalmeijer, et al. 2003. *In vitro* and *in vivo* degradation of biomimetic octacalcium phosphate and carbonate apatite coatings on titanium implants. *Journal of Biomedical Materials Research. Part A* 64(2): 378–387.

F. Barrere, C. M. van der Valk, G. Meijer, et al. 2003. Osteointegration of biomimetic apatite coating applied onto dense and porous metal implants in femurs of goats. *Journal of Biomedical Materials Research. Part B, Applied Biomaterials* 67(1): 655–665.

C. Beevers and D. McIntyre. 1946. The atomic structure of fluorapatite and its relation to that of tooth and bone mineral. *Mineralogical Magazine* 27: 254–259.

J. E. Biemond, T. S. Eufrasio, G. Hannink, et al. 2011. Assessment of bone ingrowth potential of biomimetic hydroxyapatite and brushite coated porous E-beam structures. *Journal of Materials Science. Materials in Medicine* 22(4): 917–925.

A. Bigi, E. Boanini, S. Panzavolta, et al. 2000. Biomimetic growth of hydroxyapatite on gelatin films doped with sodium polyacrylate. *Biomacromolecules* 1(4): 752–756.

E. Boanini, M. Gazzano, and A. Bigi. 2010. Ionic substitutions in calcium phosphates synthesized at low temperature. *Acta Biomaterialia* 6(6): 1882–1894.

S. Bodhak, S. Bose, and A. Bandyopadhyay. 2011. Influence of MgO, SrO, and ZnO dopants on electro-thermal polarization behavior and *in vitro* biological properties of hydroxyapatite ceramics. *Journal of the American Ceramic Society* 94(4): 1281–1288.

E. Bonnelye, A. Chabadel, F. Saltel, et al. 2008. Dual effect of strontium ranelate: Stimulation of osteoblast differentiation and inhibition of osteoclast formation and resorption *in vitro*. *Bone* 42(1): 129–138.

S. Bose, S. Tarafder, S. S. Banerjee, et al. 2011. Understanding *in vivo* response and mechanical property variation in MgO, SrO and SiO(2) doped beta-TCP. *Bone* 48(6): 1282–1290.

L. Boyer, J. Carpena, and J. L. Lacout. 1997. Synthesis of phosphate-silicate apatites at atmospheric pressure. *Solid State Ionics* 95(1–2): 121–129.

D. M. Brunette and B. Chehroudi. 1999. The effects of the surface topography of micromachined titanium substrata on cell behavior *in vitro* and *in vivo*. *Journal of Biomechanical Engineering-Transactions of the ASME* 121(1): 49–57.

J. Buehler, P. Chappuis, J. L. Saffar, et al. 2001. Strontium ranelate inhibits bone resorption while maintaining bone formation in alveolar bone in monkeys (*Macaca fascicularis*). *Bone* 29(2): 176–179.

W. P. Cao and L. L. Hench. 1996. Bioactive materials. *Ceramics International* 22(6): 493–507.

S. B. Cho, K. Nakanishi, T. Kokubo, et al. 1995. Dependence of apatite formation on silica-gel on its structure—Effect of heat-treatment. *Journal of the American Ceramic Society* 78(7): 1769–1774.

J. Christoffersen, M. R. Christoffersen, N. Kolthoff, et al. 1997. Effects of strontium ions on growth and dissolution of hydroxyapatite and on bone mineral detection. *Bone* 20(1): 47–54.

H. Colfen and S. Mann. 2003. Higher-order organization by mesoscale self-assembly and transformation of hybrid nanostructures. *Angewandte Chemie-International Edition* 42(21): 2350–2365.

G. Daculsi, R. Z. Legeros, M. Heughebaert, et al. 1990. Formation of carbonate-apatite crystals after implantation of calcium-phosphate ceramics. *Calcified Tissue International* 46(1): 20–27.

J. E. Davies. 2003. Understanding peri-implant endosseous healing. *Journal of Dental Education* 67(8): 932–949.

K. de Groot. 1989. Hydroxyapatite coated implants. *Journal of Biomedical Materials Research* 23: 1367–1371.

K. de Groot, R. Geesink, C. P. A. T. Klein, et al. 1987. Plasma sprayed coatings of hydroxylapatite. *Journal of Biomedical Materials Research* 21(12): 1375–1381.

S. Deville. 2008. Freeze-casting of porous ceramics: A review of current achievements and issues. *Advanced Engineering Materials* 10(3): 155–169.

S. Deville. 2010. Freeze-casting of porous biomaterials: Structure, properties and opportunities. *Materials* 3(3): 1913–1927.

S. Deville, E. Saiz, R. K. Nalla, et al. 2006. Freezing as a path to build complex composites. *Science* 311(5760): 515–518.

S. Deville, E. Saiz, and A. P. Tomsia. 2006. Freeze casting of hydroxyapatite scaffolds for bone tissue engineering. *Biomaterials* 27(32): 5480–5489.

W. J. Dhert, C. P. Klein, J. A. Jansen, et al. 1993. A histological and histomorphometrical investigation of fluorapatite, magnesiumwhitlockite, and hydroxylapatite plasma-sprayed coatings in goats. *Journal of Biomedical Materials Research* 27(1): 127–138.

S. Dorozhkin. 2009. Calcium orthophosphates in nature, biology and medicine. *Materials* 2(2): 399–498.

S. V. Dorozhkin. 2010. Bioceramics of calcium orthophosphates. *Biomaterials* 31(7): 1465–1485.

S. V. Dorozhkin and M. Epple. 2002. Biological and medical significance of calcium phosphates. *Angewandte Chemie* 41(17): 3130–3146.

P. Ducheyne, S. Radin, and L. King. 1993. The effect of calcium phosphate ceramic composition and structure *in vitro* behavior. I. Dissolution. *Journal of Biomedical Materials Research* 27: 25–34.

W. R. Earle. 1943. Production of malignancy *in vitro*. IV. The mouse fibroblast cultures and changes seen in the living cells. *Journal of the National Cancer Institute* 4: 165–212.

P. Frayssinet, J. L. Trouillet, N. Rouquet, et al. 1993. Osseointegration of macroporous calcium-phosphate ceramics having a different chemical-composition. *Biomaterials* 14(6): 423–429.

Q. Fu, M. N. Rahaman, F. Dogan, et al. 2008a. Freeze casting of porous hydroxy-apatite scaffolds. I. Processing and general microstructure. *Journal of Biomedical Materials Research Part Q.*

Fu, M. N. Rahaman, F. Dogan, et al. 2008b. Freeze casting of porous hydroxyapa-tite scaffolds. II. Sintering, microstructure, and mechanical behavior. *Journal of Biomedical Materials Research Part B: Applied Biomaterials* 86B(2): 514–522.

Q. Fu, M. N. Rahaman, F. Dogan, et al. 2008c. Freeze-cast hydroxyapatite scaffolds for bone tissue engineering applications. *Biomedical Materials* 3(2): 025005.

T. Fukasawa, Z. Y. Deng, M. Ando, et al. 2001. Pore structure of porous ceramics synthesized from water-based slurry by freeze-dry process. *Journal of Materials Science* 36(10): 2523–2527.

J. E. Gamble. 1967. *Chemical anatomy, physiology and pathology of extracellular fluid.* Cambridge, MA, Harvard University Press.

I. R. Gibson, S. M. Best, and W. Bonfield. 1999. Chemical characterization of sili-con-substituted hydroxyapatite. *Journal of Biomedical Materials Research* 44(4): 422–428.

I. R. Gibson, S. M. Best, and W. Bonfield. 2002. Effect of silicon substitution on the sintering and microstructure of hydroxyapatite. *Journal of the American Ceramic Society* 85(11): 2771–2777.

K. Gotfredsen, A. Wennerberg, C. Johansson, et al. 1995. Anchorage of TiO_2-blasted, HA-coated, and machined implants: An experimental study with rabbits. *Journal of Biomedical Materials Research* 29(10): 1223–1231.

Z. G. Gou and J. Chang. 2004. Synthesis and *in vitro* bioactivity of dicalcium silicate powders. *Journal of the European Ceramic Society* 24(1): 93–99.

Z. G. Gou, J. Chang, W. Y. Zhai, et al. 2005. Study on the self-setting property and the *in vitro* bioactivity of beta-Ca2SiO4. *Journal of Biomedical Materials Research Part B—Applied Biomaterials* 73B(2): 244–251.

Z. R. Gou, J. Chang, J. H. Gao, et al. 2004. *In vitro* bioactivity and dissolution of $Ca_2(SiO_3)(OH)_2$ and beta-Ca_2SiO_4 fibers. *Journal of the European Ceramic Society* 24(13): 3491–3497.

J. R. Groza. 1999. Nanosintering. *Nanostructured Materials* 12(5–8): 987–992.

P. Habibovic, F. Barrere, C. A. van Blitterswijk, et al. 2002. Biomimetic hydroxyapatite coating on metal implants. *Journal of the American Ceramic Society* 85(3): 517–522.

P. Habibovic, J. Li, C. M. van der Valk, et al. 2005. Biological performance of uncoated and octacalcium phosphate-coated Ti6Al4V. *Biomaterials* 26(1): 23–36.

J. H. Hanks. 1975. Hanks' balanced salt solution and pH control. *Methods in Cell Science* 1(1): 3–4.

J. H. Hanks and R. E. Wallace. 1949. Relation of oxygen and temperature in the pres-ervation of tissues by refrigeration. *Proceedings of the Society for Experimental Biology and Medicine* 71(2): 196–200.

L. Hench. 1991. Bioceramics: From concept to clinic. *Journal of American Ceramic Society* 74: 1487–1510.

L. Hermansson. 2011. Nanostructural chemically bonded Ca-aluminate based bioc-eramics. *Biomaterials—Physics and Chemistry.* InTech.

L. Hermansson, L. Kraft, K. Lindqvist, et al. 2008. Flexural strength measurement of ceramic dental restorative materials. *Key Engineering Materials* 361–363: 873–876.

K. A. Hing, S. M. Best, and W. Bonfield. 1999. Characterization of porous hydroxyapa-tite. *Journal of Materials Science: Materials in Medicine* 10(3): 135–145.

K. Hyakuna, T. Yamamuro, Y. Kotoura, et al. 1990. Surface-reactions of calcium-phosphate ceramics to various solutions. *Journal of Biomedical Materials Research* 24(4): 471–488.

Y. Ito. 1999. Surface micropatterning to regulate cell functions. *Biomaterials* 20(23–24): 2333–2342.

I. Izquierdo-Barba, D. Arcos, Y. Sakamoto, et al. 2008. High-performance mesoporous bioceramics mimicking bone mineralization. *Chemistry of Materials* 20(9): 3191–3198.

I. Izquierdo-Barba, M. Colilla, and M. Vallet-Regi. 2008. Nanostructured mesoporous silicas for bone tissue regeneration. *Journal of Nanomaterials* 106970.

S. R. Kim, J. H. Lee, Y. T. Kim, et al. 2003. Synthesis of Si, Mg substituted hydroxyapatites and their sintering behaviors. *Biomaterials* 24(8): 1389–1398.

C. P. Klein, P. Patka, J. G. Wolke, et al. 1994. Long-term *in vivo* study of plasma-sprayed coatings on titanium alloys of tetracalcium phosphate, hydroxyapatite and alpha-tricalcium phosphate. *Biomaterials* 15(2): 146–150.

T. Kokubo. 1991. Bioactive glass-ceramics—Properties and applications. *Biomaterials* 12(2): 155–163.

T. Kokubo, H. Kushitani, C. Ohtsuki, et al. 1992. Chemical-reaction of bioactive glass and glass-ceramics with a simulated body-fluid. *Journal of Materials Science: Materials in Medicine* 3(2): 79–83.

T. Kokubo, H. Kushitani, S. Sakka, et al. 1990. Solutions able to reproduce *in vivo* surface-structure changes in bioactive glass-ceramic A-W. *Journal of Biomedical Materials Research* 24(6): 721–734.

L. B. Kong, J. Ma, and F. Boey. 2002. Nanosized hydroxyapatite powders derived from coprecipitation process. *Journal of Materials Science* 37(6): 1131–1134.

T. A. Kuriakose, S. N. Kalkura, M. Palanichamy, et al. 2004. Synthesis of stoichiometric nano crystalline hydroxyapatite by ethanol-based sol–gel technique at low temperature. *Journal of Crystal Growth* 263(1–4): 517–523.

S. Kuroda, A. S. Virdi, P. Li, et al. 2004. A low-temperature biomimetic calcium phosphate surface enhances early implant fixation in a rat model. *Journal of Biomedical Materials Research. Part A* 70(1): 66–73.

E. Landi, G. Logroscino, L. Proietti, et al. 2008. Biomimetic Mg-substituted hydroxyapatite: From synthesis to *in vivo* behaviour. *Journal of Materials Science: Materials in Medicine* 19(1): 239–247.

E. Landi, F. Valentini, and A. Tampieri. 2008. Porous hydroxyapatite/gelatine scaffolds with ice-designed channel-like porosity for biomedical applications. *Acta Biomaterialia* 4(6): 1620–1626.

S. Langstaff, M. Sayer, T. J. N. Smith, et al. 1999. Resorbable bioceramics based on stabilized calcium phosphates. Part I: Rational design, sample preparation and material characterization. *Biomaterials* 20(18): 1727–1741.

K. Y. Lee, M. Park, H. M. Kim, et al. 2006. Ceramic bioactivity: Progresses, challenges and perspectives. *Biomedical Materials* 1(2): R31–R37.

S. Leeuwenburgh, P. Layrolle, F. Barrere, et al. 2001. Osteoclastic resorption of biomimetic calcium phosphate coatings *in vitro*. *Journal of Biomedical Materials Research* 56(2): 208–215.

R. Z. LeGeros. 1993. Biodegradation and bioresorption of calcium phosphate ceramics. *Clinical Materials* 14(1): 65–88.

J. E. Lemons, N. Z. Ramsay, and E. K. Chamoun. 1988. Dental implant device retrievals. SFB Symposium on Retrieval and Analysis of Surgical Implant and Biomaterials, Snowbird, Utah.

I. B. Leonor, A. Ito, K. Onuma, et al. 2002. In situ study of partially crystallized bio-glass and hydroxylapatite *in vitro* bioactivity using atomic force microscopy. *Journal of Biomedical Materials Research* 62(1): 82–88.

P. J. Li, C. Ohtsuki, T. Kokubo, et al. 1994. The role of hydrated silica, titania, and alumina in inducing apatite on implants. *Journal of Biomedical Materials Research* 28(1): 7–15.

X. Li, J. L. Shi, Y. F. Zhu, et al. 2007. A template route to the preparation of mesoporous amorphous calcium silicate with high *in vitro* bone-forming bioactivity. *Journal of Biomedical Materials Research Part B–Applied Biomaterials* 83B(2): 431–439.

X. Li, H. Yasuda, and Y. Umakoshi. 2006. Bioactive ceramic composites sintered from hydroxyapatite and silica at 1200 degrees C: Preparation, microstructures and *in vitro* bone-like layer growth. *Journal of Materials Science: Materials in Medicine* 17(6): 573–581.

K. L. Lin, W. Y. Zhai, S. Y. Ni, et al. 2005. Study of the mechanical property and *in vitro* biocompatibility of $CaSiO_3$ ceramics. *Ceramics International* 31(2): 323–326.

C. Lindahl, P. Borchardt, J. Lausmaa, et al. 2010. Studies of early growth mechanisms of hydroxyapatite on single crystalline rutile: A model system for bioactive sur-faces. *Journal of Materials Science: Materials in Medicine* 21: 2743–2749.

F. Lindberg, J. Heinrichs, F. Ericson, et al. 2008. Hydroxylapatite growth of single-crystal rutile substrates. *Biomaterials* 29: 3317–3323.

C. M. Muller-Mai, C. Voigt, and U. Gross. 1990. Incorporation and degradation of hydroxyapatite implants of different surface roughness and surface structure in bone. *Scanning Microscopy* 4(3): 613–622; discussion 622–614.

M. K. Narbat, F. Orang, M. S. Hashtjin, and A. Goudarzi. 2006. Fabrication of porous hydroxyapatite-gelatin composite scaffolds for bone tissue engineering. *Iran. Biomedical Journal* 10: 215–223.

M. D. O'Donnell, Y. Fredholm, A. de Rouffignac, et al. 2008. Structural analysis of a series of strontium-substituted apatites. *Acta Biomaterialia* 4(5): 1455–1464.

C. Ohtsuki, Y. Aoki, T. Kokubo, et al. 1995. Transmission electron microscopic obser-vation of glass-ceramic A-W and apatite layer formed on its surface in a simu-lated body fluid. *Journal of the Ceramic Society Japan* 103(5): 449–454.

C. Ohtsuki, H. Kushitani, T. Kokubo, et al. 1991. Apatite formation on the surface of ceravital-type glass-ceramic in the body. *Journal of Biomedical Materials Research* 25(11): 1363–1370.

M. Okazaki, J. Takahashi, H. Kimura, et al. 1982. Crystallinity, solubility, and disso-lution rate behavior of fluoridated CO3 apatites. *Journal of Biomedical Materials Research* 16(6): 851–860.

A. Oyane, K. Onuma, A. Ito, et al. 2003. Formation and growth of clusters in conven-tional and new kinds of simulated body fluids. *Journal of Biomedical Materials Research. Part A* 64(2): 339–348.

H. B. Pan, Z. Y. Li, W. M. Lam, et al. 2009. Solubility of strontium-substituted apatite by solid titration. *Acta Biomaterialia* 5(5): 1678–1685.

J. D. Pasteris, B. Wopenka, and E. Valsami-Jones. 2008. Bone and tooth mineralization: Why apatite? *Elements* 4: 97–104.

A. M. Pietak, J. W. Reid, M. J. Stott, et al. 2007. Silicon substitution in the calcium phosphate bioceramics. *Biomaterials* 28(28): 4023–4032.

A. E. Porter, S. M. Best, and W. Bonfield. 2004. Ultrastructural comparison of hydroxy-apatite and silicon-substituted hydroxyapatite for biomedical applications. *Journal of Biomedical Materials Research Part A* 68A(1): 133–141.

M. N. Rahaman and Q. Fu. 2008. Manipulation of porous bioceramic microstructures by freezing of suspensions containing binary mixtures of solvents. *Journal of the American Ceramic Society* 91(12): 4137–4140.

R. R. Rao, H. N. Roopa, and T. S. Kannan. 1997. Solid state synthesis and thermal stability of HAP and HAP–β-TCP composite ceramic powders. *Journal of Materials Science: Materials in Medicine* 8(8): 511–518.

B. D. Ratner and S. J. Bryant. 2004. Biomaterials: Where we have been and where we are going. *Annual Review of Biomedical Engineering* 6: 41–75.

S. Raynaud, E. Champion, and D. Bernache-Assollant. 2002. Calcium phosphate apatites with variable Ca/P atomic ratio II. Calcination and sintering. *Biomaterials* 23(4): 1073–1080.

S. Raynaud, E. Champion, J. P. Lafon, et al. 2002. Calcium phosphate apatites with variable Ca/P atomic ratio III. Mechanical properties and degradation in solution of hot pressed ceramics. *Biomaterials* 23(4): 1081–1089.

J. W. Reid, A. Pietak, M. Sayer, et al. 2005. Phase formation and evolution in the silicon substituted tricalcium phosphate/apatite system. *Biomaterials* 26(16): 2887–2897.

F. Ren, R. Xin, X. Ge, et al. 2009. Characterization and structural analysis of zinc-substituted hydroxyapatites. *Acta Biomaterialia* 5(8): 3141–3149.

G. Renaudin, P. Laquerriere, Y. Filinchuk, et al. 2008. Structural characterization of sol-gel derived Sr-substituted calcium phosphates with anti-osteoporotic and anti-inflammatory properties. *Journal of Materials Chemistry* 18(30): 3593–3600.

S. Ringer. 1883. A further contribution regarding the influence of the different constituents of the blood on the contraction of the heart. *Journal of Physiology* 4(1): 29–42.

U. Ripamonti, J. Crooks, L. Khoali, et al. 2009. The induction of bone formation by coral-derived calcium carbonate/hydroxyapatite constructs. *Biomaterials* 30(7): 1428–1439.

D. M. Roy and S. K. Linnehan. 1974. Hydroxyapatite formed from coral skeletal carbonate by hydrothermal exchange. *Nature* 247(5438): 220–222.

M. Sayer, A. D. Stratilatov, J. Reid, et al. 2003. Structure and composition of silicon-stabilized tricalcium phosphate. *Biomaterials* 24(3): 369–382.

H. K. Schmidt. 2000. Nanoparticles for ceramic and nanocomposite processing. *Molecular Crystals and Liquid Crystals* 353: 165–179.

H. Schoof, J. Apel, I. Heschel, et al. 2001. Control of pore structure and size in freeze-dried collagen sponges. *Journal of Biomedical Materials Research* 58(4): 352–357.

H. Schoof, L. Bruns, J. Apel, et al. 1998. Einfluss des Einfriervorganges auf die Porenstruktur gefriergetrockneter Kollagenschwämme. *KiLuft und Kätetechnik* 34: 247–252.

H. Schoof, L. Bruns, A. Fischer, et al. 2000. Dendritic ice morphology in unidirectionally solidified collagen suspensions. *Journal of Crystal Growth* 209(1): 122–129.

A. G. Shard and P. E. Tomlins. 2006. Biocompatibility and the efficacy of medical implants. *Regenerative Medicine* 1(6): 789–800.

E. C. Shors. 1999. Coralline bone graft substitutes. *The Orthopedic Clinics of North America* 30(4): 599–613.

A. Siddharthan, S. K. Seshadri, and T. S. S. Kumar. 2006. Influence of microwave power on nanosized hydroxyapatite particles. *Scripta Materialia* 55(2): 175–178.

S. I. Stupp and P. V. Braun. 1997. Molecular manipulation of microstructures: Biomaterials, ceramics, and semiconductors. *Science* 277(5330): 1242–1248.

A. Tampieri, G. Celotti, S. Sprio, et al. 2001. Porosity-graded hydroxyapatite ceramics to replace natural bone. *Biomaterials* 22(11): 1365–1370.

A. Tampieri, S. Sprio, A. Ruffini, et al. 2009. From wood to bone: Multi-step process to convert wood hierarchical structures into biomimetic hydroxyapatite scaffolds for bone tissue engineering. *Journal of Materials Chemistry* 19(28): 4973–4980.

D. C. Tancred, B. A. O. McCormack, and A. J. Carr. 1998. A synthetic bone implant macroscopically identical to cancellous bone. *Biomaterials* 19(24): 2303–2311.

X. L. Tang, X. F. Xiao, and R. F. Liu. 2005. Structural characterization of silicon-substituted hydroxyapatite synthesized by a hydrothermal method. *Materials Letters* 59(29–30): 3841–3846.

A. C. Tas. 2000. Synthesis of biomimetic Ca-hydroxyapatite powders at 37 degrees C in synthetic body fluids. *Biomaterials* 21(14): 1429–1438.

P. Thevenot, W. J. Hu, and L. P. Tang. 2008. Surface chemistry influences implant biocompatibility. *Current Topics in Medicinal Chemistry* 8(4): 270–280.

M. Uchida, H. M. Kim, T. Kokubo, et al. 2001. Apatite-forming ability of sodium-containing titania gels in a simulated body fluid. *Journal of the American Ceramic Society* 84(12): 2969–2974.

M. Vaahtio, T. Peltola, T. Hentunen, et al. 2006. The properties of biomimetically processed calcium phosphate on bioactive ceramics and their response on bone cells. *Journal of Materials Science-Materials in Medicine* 17(11): 1113–1125.

M. Vallet-Regi and D. Arcos. 2005. Silicon substituted hydroxyapatites. A method to upgrade calcium phosphate based implants. *Journal of Materials Chemistry* 15(15): 1509–1516.

M. Vallet-Regi and J. M. Gonzalez-Calbet. 2004. Calcium phosphates as substitution of bone tissues. *Progress in Solid State Chemistry* 32(1–2): 1–31.

D. Veljović, B. Jokić, R. Petrović, et al. 2009. Processing of dense nanostructured HAP ceramics by sintering and hot pressing. *Ceramics International* 35(4): 1407–1413.

X. Wang, Y. Li, J. Wei, et al. 2002. Development of biomimetic nano-hydroxyapatite/poly(hexamethylene adipamide) composites. *Biomaterials* 23(24): 4787–4791.

T. J. Webster, C. Ergun, R. H. Doremus, et al. 2001. Enhanced osteoclast-like cell functions on nanophase ceramics. *Biomaterials* 22(11): 1327–1333.

U. G. K. Wegst, M. Schecter, A. E. Donius, et al. 2010. Biomaterials by freeze casting. *Philosophical Transactions of the Royal Society A: Mathematical, Physical and Engineering Sciences* 368(1917): 2099–2121.

M. Wei, M. Uchida, H. M. Kim, et al. 2002. Apatite-forming ability of CaO-containing titania. *Biomaterials* 23(1): 167–172.

H. B. Wen, J. G. C. Wolke, J. R. deWijn, et al. 1997. Fast precipitation of calcium phosphate layers on titanium induced by simple chemical treatments. *Biomaterials* 18(22): 1471–1478.

R. A. White, J. N. Weber, and E. W. White. 1972. Replamineform: A new process for preparing porous ceramic, metal, and polymer prosthetic materials. *Science* 176(4037): 922–924.

A. K. A. R. Wijenayaka, C. B. Colby, G. J. Atkins, et al. 2009. Biomimetic hydroxyapatite coating on glass coverslips for the assay of osteoclast activity *in vitro*. *Journal of Materials Science-Materials in Medicine* 20(7): 1467–1473.

C. J. Wilson, R. E. Clegg, D. I. Leavesley, et al. 2005. Mediation of biomaterial-cell interactions by adsorbed proteins: A review. *Tissue Engineering* 11(1–2): 1–18.

C. T. Wu and J. Chang. 2004. Synthesis and apatite-formation ability of akermanite. *Materials Letters* 58(19): 2415–2417.

W. Xia, C. Lindahl, J. Lausmaa, et al. 2010. Biomineralized strontium-substituted apatite/titanium dioxide coating on titanium surfaces. *Acta Biomaterialia* 6(4): 1591–1600.

W. Xia, C. Lindahl, C. Persson, et al. 2010. Changes of surface composition and morphology after incorporation of ions into biomimetic apatite coating. *Journal of Biomaterials and Nanobiotechnology* 1: 7–16.

J. L. Xu, K. A. Khor, Y. W. Gu, et al. 2005. Radio frequency (rf) plasma spheroidized HA powders: Powder characterization and spark plasma sintering behavior. *Biomaterials* 26(15): 2197–2207.

E. Zhang, C. M. Zou, and G. N. Yu. 2009. Surface microstructure and cell biocompatibility of silicon-substituted hydroxyapatite coating on titanium substrate prepared by a biomimetic process. *Materials Science & Engineering C-Biomimetic and Supramolecular Systems* 29(1): 298–305.

J. Zhao, L. Y. Guo, X. B. Yang, et al. 2008. Preparation of bioactive porous HA/PCL composite scaffolds. *Applied Surface Science* 255(5, Part 2): 2942–2946.

W. Y. Zhao and J. Chang. 2004. Sol-gel synthesis and *in vitro* bioactivity of tricalcium silicate powders. *Materials Letters* 58(19): 2350–2353.

W. Y. Zhao and J. Chang. 2005. Preparation and characterization of novel tricalcium silicate bioceramics. *Journal of Biomedical Materials Research Part A* 73A(1): 86–89.

6

Preparation and Mechanism of Novel Bioceramics with Controllable Morphology and Crystal Growth

Kaili Lin and Jiang Chang

CONTENTS

6.1 Introduction

The ceramics used for repair and reconstruction of diseased or damaged parts of the muscular-skeletal system are termed as bioceramics (Hench 1991). The bioceramics can be classified into bioinert, resorbable, and bioactive ceramic materials. They can be used in dense, porous, granules, particles, and coatings, and also used as the component for composites. The materials based on calcium phosphate (Ca-P), such as hydroxyapatite [$Ca_{10}(PO_4)_6(OH)_2$, HAp], β-tricalcium phosphate [$β-Ca_3(PO_4)_2$, β-TCP], octacalcium phosphate [$Ca_8H_2(PO_4)_6 \cdot 5H_2O$, OCP], and calcium phosphate cement (CPC) are widely used in biomedical fields due to their excellent biocompatibility, osteoconductive properties, and similarity to the inorganic component of human beings (Lin et al. 2012). In recent decades, another kind of bioceramics, materials based on calcium silicate (Ca-Si), such as bioglass, glass-ceramics, and calcium silicate ceramics, are developed due to their bioactivity and biodegradability. The bioceramics can be also used as a drug or gene delivery system.

The morphology, structure, and size of bioceramic crystal and aggregates influence the performance in their applications. For example, rodlike, wire-like, and sheetlike particles have a stronger molecular adsorption property due to the increased surface area, whereas rodlike, whiskerlike, wirelike, and sheetlike inorganics can be used as mechanical reinforcement to fabricate biocomposites because of their excellent mechanical properties. The nanostructured porous or hollow materials with hierarchical architectures can be used as a drug or gene delivery system because of their high drug loading and favorable controllable release properties.

Material with initiative stimulation capacity in tissue regeneration is the major character for next-generation biomaterials (Hench and Polak 2002). However, up to now most of the bioceramics, including the traditional Ca-P–based bioceramics, lack the ability to stimulate the formation of new bone, which hindered their clinical applications (Tabrizi et al. 2009). Several attempts have been made to solve this problem, such as loading bioactive growth factors (Notodihardjo et al. 2012), surface morphology and topology design (Wong et al. 1995; Kuboki et al. 1998; Ribeiro et al. 2010), and the incorporation of the functional trace elements (Lin, Chang, Liu, et al. 2011; Lin, Zhou, et al. 2011). Recently, the improvement of the biological responses via surface morphology and topology design has aroused great interest.

Various strategies have been developed to fabricate bioceramic materials with different morphologies, such as platelike (or sheetlike); needles, rods, whiskers, and fibers; mesostructures; spheres; core-shell structures; and three-dimensional (3D) hierarchical architectures from nano- to microscales. The most common methods to control the morphologies are based on the template-directing strategy. The templates could be "soft materials," such as surfactants, polymers, and biomoleculars, or "hard materials," such as calcium carbonates, calcium phosphate, and calcium silicates. The mechanisms behind the template-directing processes are different. Another interesting strategy of morphology control is the self-assembly without using any templates, such as mineralization, which are known as wet chemical routes.

In this chapter, the recent developments of the preparation and mechanism of novel bioceramics with controllable morphologies are summarized. Due to the limitation on length, most examples focus on Ca-P–based materials.

6.2 Preparation and Mechanism of Novel Bioceramics with Controllable Morphologies and Crystal Growth

6.2.1 Crystal Growth and Morphology Control of Bioceramic Particles

Many methods have been applied to synthesize bioceramic particles with different morphologies, crystal sizes, and chemical components. These methods can be mainly divided into several kinds of strategies, such as solid-state reaction, wet chemical process, precursor transformation method (also regarded as "hard template" method), and physical approach. The solid-state reactions usually result in agglomerate granulars in micro or submicro size level even after ball milling because of the overly high reaction temperatures. Moreover, the sintering ability of such powders is usually low and ultimately results in lower mechanical properties of the sintered matrixes (Byrappa 2001). The wet technique includes chemical precipitation (Lin, Pan et al. 2009), sol-gel (Liu et al. 2001), hydrothermal (Byrappa and Adschiri 2007), microemulsion (Lim et al. 1997), and solution reaction with the assistance of the microwave (Li et al. 2011), ultrasound (Bang and Suslick 2010), and ultraviolet (Nishikawa 2003) irradiation. As for the wet chemical process, there are various surfactants, organic solvent, or molecular template-directing reagents, such as ethylene diamine tetraacetic acid (EDTA), sodium ethylene diamine tetraacetic acid (Na_2EDTA), N, N-dimethylformamide (DMF), hexadecyltrimethylammonium bromide (CTAB), bis(2-ethylhexyl) sulfosuccinate (AOT), sodium dodecyl sulfate (SDS), isopropyl alcohol, hexane, glycine, formamide, examethylenetetramine, amino acids, protein, and monosaccharide, which can be regarded as "soft template," and are widely used to control the morphologies, crystal growth direction, crystal sizes, stoichiometry, ion substitution, and the crystallinity of the bioceramic particles required for specific applications in the wet chemical process. The transformation of the solid precursors into products with designed morphologies and chemical compositions through oxidation-reduction, replacement reaction, hydrolysis and hydrothermal treatment, is another choice. This approach can be regarded as the "hard template" method. The freeze-drying, mechanochemical method, electrospinning, and spray pyrolysis are widely applied as the physical approaches to fabricate bioceramics with specific morphologies.

In general, natural bioinorganic apatite crystals in bone and tooth, and the chemical precipitated hydroxyapatite (HAp) particles are always needle-, rod-, fiber-, or platelike in thin thickness because of the preferred orientation growth of HAp crystal along the c-axis (Su et al. 2003; Dorozhkin 2007). Another explanation is that the plate-shaped OCP is the precursor of HAp crystals, which grow along the OCP transition phase (Weiner and Wagner 1998). The energy barrier for nucleating HAp is higher than that of OCP since the surface energy of OCP is lower than that of HAp. Therefore, the HAp prefers to grow along the OCP layer and results in a platelike shape. Viswanath and Ravishankar (2008) developed a general methodology to illustrate the evidence for the formation of the platelike HAp. Their results showed that the lowest energy surface is the prism plane (1 0 0). They strongly recommend that the platelike shape of OCP and HAp is mainly due to the chemical driving force at which OCP or HAp form falls in the layer-by-layer growth zone. The relatively low temperature and neutral pH value favor the growth of two-dimensional nanostructures associated with a low chemical driving force (Viswanath and Ravishankar 2008). The preferential crystal growth habit of HAp nanocrystals along the (0 0 0 1) face was observed by real-time phase shift interferometry and atomic force microscopy (AFM) technology in simulated body fluid (SBF) condition (Onuma et al. 1998).

Furthermore, the physicochemical properties of the bioceramic particles are also remarkably influenced by the reaction conditions, such as the type of raw materials and mixing types, molar ratio, concentrations, molecule and ion additives, pH value, temperatures, and aging time. Zhang and Lu (2007) found that the modes of reaction, namely, quick mixing and drop-let adding, was in favor of spherical and rodlike morphologies, respectively. When the $(NH_4)_2HPO_4$ water solution was quickly added into the $Ca(NO_3)_2$ ethanol solution, the spherical HAp particles were obtained. This might be attributed to the fast mixing process, which enables homogenous nucleation of HAp particles and hence restrains the preferential growth of HAp, and subsequently a uniform chemical reaction occurred. The temperatures can also significantly influence the morphology and size of the bioceramic materials (Kumar et al. 2004). In the synthesis of HAp materials via wet chemical precipitation route using $Ca(OH)_2$ and H_3PO_4 as raw materials, needlelike nanoparticles with a high aspect ratio were obtained at 40°C, while spherical particles were obtained with the increase of the precipitation temperature to 100°C. The analysis indicated that the supersaturating level of the reactants, especially the concentration of Ca^{2+} ions, played a major role in the precipitate morphology in the acid-base reaction system.

Iijima et al. (1996) investigated the role of F^- ions on apatite/OCP/apatite structure formation mechanism. The following conclusion was obtained: When F^- ions existed in solution where OCP precipitated, F^- ions enabled sequential precipitation of a low-substituted FHAp on an OCP template, which incorporated a small amount of F^-. A small amount of F^--containing OCP, or a surface reaction layer of OCP that has accumulated a small amount of F^-, was suggested to be formed. The roles of F^- in the HAp/OCP/HAp structure formation were hypothesized as the reduction of the growth rate and the critical thickness in the a-axis direction of OCP, the enhancement of hydrolysis of OCP, and the activation of the growth of FHAp, which resulted in thinner OCP lamella and thicker apatite lamella in the a-axis direction with an increase in the F^- concentration. The study showed that the carbonate substitution caused a reduction in crystallinity and changed in the shape from needlelike to rod-shaped to platelike (equiaxed) apatite crystals (Robinson 1952; Bocciarelli 1970; Weiner et al. 1991). The presence of chloride ions in the crystallizing medium favored the formation of platelike HAp crystals when a large amount of precipitation took place (Koutsoukos and Nancollas 1981). Kanchana and Sekar (2010) also found that the addition of NaF significantly reduced the growth rate and the yield of HAp in the diffusion gel method using sodium meta silicate (SMS) gel. At the same time, the microstructural morphology of the synthesized HAp changed from fibrous to granular structure due to fluoride substitution, and the crystallinity of the HAp increased with fluoride substitution. The study of Huang et al. (2012) suggested that the Eu ion inhibited the crystal growth in the c-axis direction and decreased the ratio of width to length of HAp crystals. The possible reason was that the replacement of Ca by Eu inhibited the crystal growth along the active plane.

Apart from the influence of ion additives on the resultant morphologies, the organics, including the surfactants, polymers, and proteins, are another choice to be used as templates in solution systems to control the shape and size of HAp crystals. The previous studies revealed that organic molecules could be absorbed on the crystal surfaces and regulated the morphologies of the products (Chander and Fuerstenau 1984; Burke et al. 2000; Spanos et al. 2001). Bose and Saha (2003) synthesized spherical-like nanocrystalline HAp powder with particle diameters of 30 and 50 nm using the reverse microemulsion method, cyclohexane as the oil phase, and a mixture of poly(oxyethylene) nonylphenol ether and poly(oxyethylene) nonylphenol ether (NP-12) as the surfactant phase. Cai et al. (2007) using hexadecyl (cetyl) trimethylammonium bromide (CTAB) as the efficient reagent to regulate the morphologies and sizes of HAp nanocrystals. The particle sizes of the HAp spheres can be facilely regulated via changing the concentration of CTAB in the calcium phosphate supersaturated solutions. Three different spherical-like HAp nanoparticles with uniform and average diameters of 20±5, 40±10, and 80±12 nm, respectively, were obtained under different concentrations of CTAB. In contrast, HAp nanoparticles grown in the absence of organic additives are typically rodlike particles with lengths of hundreds of nano-meters and widths of tens of nanometers. The glutamic acid was used as the additive to hydrothermally synthesize large-sized HAp whiskers with length of 50 to 100 μm and width of 0.5 μm in a dilute reaction solution, where the supersaturation degree with respect to HAp precipitation was low. It was found that the different adsorption and desorption behaviors of Glu occurred on (1 0 0) and (0 0 1) planes of HAp and resulted in the crystal growth along the c-axis (Li et al. 2010). The peptides possess well affinity with HAp, which has been developed to modulate the morphology and size of the HAp crystals. The research of Diegmueller et al. (2009) showed the type of charged peptide, peptidic molecular weight, and concentration on the morphology and size of the HAp crystals. In their study, the poly-L-argi-nine (Arg), poly-L-aspartate (Asp), poly-L-glutamate (Glu), and poly-L-lysine (Lys) with different molecular weight were synthesized, and then were used as the biomimetic analogs of noncollagenous proteins to investigate their influence on mineral nucleation and growth kinetics, and crystal morphol-ogy of HAp. The results showed that the negatively charged polymers (Asp, Glu) and higher molecular weight (HMW) had greater affinity for HAp than positively charged polymers (Arg, Lys) and lower molecular weight (LMW). The poly-L-Asp LMW, poly-L-Glu LMW, and poly-L-Lys HMW were found to significantly increase the aspect ratio in comparison to the control. Such knowledge could be useful for identifying the peptides ideal for *in vitro* or *in vivo* administration to customize crystal growth or for coating of implant surfaces for improving anchorage to neighboring bone tissue.

Yokoi et al. (2010) used polyacrylamide (PAAm) hydrogel as the template to synthesize calcium phosphate crystals. Their study showed that the concentrations of calcium and phosphate ions played an important role on the crystalline phases and morphology of the products. Several kinds of calcium phosphate crystals were precipitated at various $Ca(NO_3)_2$ concentrations (0.5~4.0 mol·dm^{-3}) or $(NH_4)_2HPO_4$ contents (3.6~21.6 mmol) in the gels. The crystalline phases were mainly determined by the $(NH_4)_2HPO_4$ content in the gels. When the $(NH_4)_2HPO_4$ content was ≥10.8mmol, HAp formed near the interfaces between $Ca(NO_3)_2$ solution and the gels, whereas OCP formed in gels with ≤10.8mmol $(NH_4)_2HPO_4$. HAp crystals were granular in form and about 200 nm in diameter, and OCP crystals were spherulitic with diameter 10~70 µm.

Simulated body fluid (SBF) is a popular solution applied to synthesize HAp material. Usually the two key parameters of pH value and media temperature are used to adjust the morphologies of HAp crystals (Kobayashi et al. 2012). When the pH value of SBF was adjusted to 6.5, the phosphate-rich HAp nanocrystals with needlelike shape precipitated at 38°C. With the increase of the pH value to 7.0, the sheetlike nanostructures with (1 1 0) surfaces were obtained. With the increase of the SBF temperature to 160°C, HAp crystals with rodlike and platelike shapes in micrometer size were prepared at pH value of 7.0 and 7.4, respectively. It is assumed that the adsorption of phosphates to the specific faces inhibits the growth of HAp crystals and changes the morphology to low-dimensional forms. Bouyer et al. (2000) also suggested that the morphology and size of HAp crystals were sensitive to the reaction temperature. A critical temperature of 60°C could be used to define the HAp crystals in monocrystalline or polycrystalline in the synthesis process. The anisotropic growth of HAp crystals on the template of bombyx mori silk fibroin (SF) films in 1.5 times SBF at 37°C was also researched by Li et al. (2008). The result indicated that the HAp crystalline properties would be controlled effectively by the positions and density of carboxyl groups, C = O, and amino groups on the surface of SF films, which provided a mimicking biomineralization process to fabricate the polymer–apatite composites.

The hydrothermal treatment can enhance the apatite crystal growth in the c-axis and also increases the crystallinity of the products (Lin, Chang, Cheng, et al. 2007), and the similar phenomenon has also been observed in Ca-Si–based bioceramic particles (Lin et al. 2006; Lin, Chang, Chen, et al. 2007; Lin, Chang, and Cheng 2007). The HAp (Lin, Chang, Cheng et al. 2007), tobermorite (Lin, Chang, and Cheng 2007), nanowires, xonotlite, and wollastonite nanowires (Lin et al. 2006) can be synthesized via hydrothermal or hydrothermal microemulsion processes. Viswanath and Ravishankar (2008) hydrothermal prepared plate-shaped OCP and HAp crystals at the low pH value of 6. The fundamental reason behind the plate-shaped morphology control for HAp crystals was revealed. They believed that the plate-shaped OCP and HAp was mainly due to the fact that the chemical driving force at which OCP or HAp forms falls in the layer-by-layer growth zone. Furthermore, the physiological conditions existing during bone biomineralization (pH and temperature) are associated

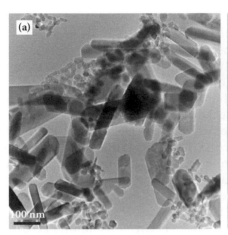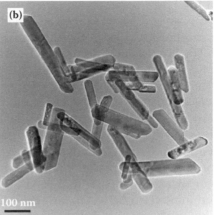

FIGURE 6.1
FETEM images of the HAp nanoparticles obtained via (a) hydrothermal treatment and (b) hydrothermal-microemulsion process at 180°C for 18 h.

with a low chemical driving force at low pH condition promoting the formation of two-dimensional (2D) nanostructures in the case of apatite presented in the bones. The similar phenomenon was observed in the formation of HAp at low temperatures by decomposition of a Ca-EDTA complex (Arce et al. 2004).

Though the hydrothermal technique usually gives HAp powders with a high degree of crystallinity and with a Ca/P ratio close to the stoichiometric value (Byrappa 2001), the obtained powders are usual in agglomeration and the size distribution is wide (Figure 6.1a). Therefore, the size distribution of the products cannot be well controlled using a normal hydrothermal method. To overcome these limitations, microemulsion and hydrothermal-microemulsion methods have been developed to synthesize inorganic materials with uniform particle morphology, size, and size distribution. Especially for the newly developed hydrothermal-microemulsion technique, which is considered as an effective, convenient, and mild synthetic methodology for the synthesis of nanopowders, nanoneedles, and nanowires (Chen et al. 2003; Zhang and Gao 2003; Lin et al. 2006). The microemulsion not only can serve as nanoreactors to control the particle size and size distribution in the reaction process but also inhibit the excess agglomeration of particles, since the surfactants can absorb on the particle surface when the particle size approaches that of the water (or oil) pool. In addition, the surfactants in the microemulsion can also serve as a versatile "soft template" for the synthesis of 1D nanostructural materials. Furthermore, the hydrothermal treatment can effectively increase the crystallinity of the product. We synthesized the stoichiometric single crystal HAp nanorods with monodispersion and narrow size distribution in diameter via reversed microemulsion (water dispersed in oil, W/O) of CTAB–n-pentanol–n-hexane–water under hydrothermal condition (Figure 6.1b) (Lin, Chang, Cheng et al. 2007). The homogeneity in size distribution and shape of the HAp nanorods was attributed to the W/O nanoreactors and the soft template of the surfactants, and the high crystallization of the products was attributed to the hydrothermal treatment.

The HAp nanowires with uniform width of 60 nm and length up to 1 μm have been synthesized by Wang, Lai, et al. (2006) via a solvothermal process. The results revealed that the formation of an amorphous nuclear–surfactant complex and the electrostatic field within the reverse micelles maintained the unidirectional, irreversible fusion of reverse micelles, leading to the growth of nanowires in one direction. In addition, the stable conditions without intense shearing stress where it was favorable for the formation of long and uniform HAp nanowires. Furthermore, the ratio of length to diameter increased apparently with the decrease of the initial pH values due to the interaction between surfactant CTAB molecules and reactant ions. The higher pH resulted in weakening interactions between the amino group of CTAB and the reactant PO_4^{3-}, leading to crystal nucleating in the bulk water phase without the growth-guiding of surfactant molecules (Wei et al. 2006).

The hydrothermal temperature and time, the additives and additive concentration, and the pH value of the solution may also play important roles with the size and morphology of the products. Wang, Chen, et al. (2006) hydrothermal synthesized HAp nanocrystals using CTAB as a template. TEM micrographs revealed the HAp particles ranged from spheroid to long fibers. The obtained smallest average diameter of spherical particles, ~27 nm, were achieved at 90°C, pH = 13 for 20 h. The biggest, longest fibers, more than 1125 nm and about 60 nm in width, were obtained at 150°C, pH = 9 for 20 h. In addition, the crystal size increased apparently with the increase of the hydrothermal temperatures. The morphology of the obtained particles presented a certain rule with the temperature at different pH values. A probable mechanism involved the interaction of the tetrahedral CTAB cation structure, the tetrahedral PO_4^{3-} structure, and the OH$^-$ ions at different pH values.

The homogeneous precipitation method under hydrothermal treatment has been widely used to control the morphologies of HAp using EDTA, Na_2EDTA, citrate (CA), urea, and amine as the chelated reagents or homogeneous precipitation reagents. Seo and Lee (2008) prepared HAp whiskers with a width of 0.05 μm and a length up to 3 μm through directly refluxing the typically shaped HAp powders using EDTA via the dissolution–reprecipitation process, in which the amorphous reprecipitates were initially formed in the Ca(EDTA)$^{2-}$–PO_4^{3-} mixed solution and have continuously grown into long HAp whiskers. Martins et al. (2008) observed the role of the existing CA species as calcium chelates under different pH conditions on the framework of particle nucleation and growth mechanisms. They suggested that the anisotropy of surface charge distribution, arising from the adsorption of citrate species onto preferential crystal facets, has a strong effect on the formation of particular particle shapes by oriented aggregation of the primary particles. In higher pH conditions, the adsorption of more negatively charged species was thought to increase the negative charge on particle shell, promoting particle repulsion that prevented particle agglomeration, thereby contributing to maintaining the nanometric size of the particles. Recently, the HAp microwhiskers more than 100 μm diameter were prepared using acetamide as an additive. The ultralong crystal product was attributed to the low hydrolysis rate of acetamide under hydrothermal conditions, the pH rising slowly, giving smaller numbers of nuclei, and facilitating whisker growth at a low supersaturation (Zhang and Darvell 2010a,b; Shen et al. 2012). The well-crystallized platelike β-TCP whiskers with width of 0.5 to 1 μm and length of 5 to 10 μm were synthesized from an acid solution (initial pH = 4.5) through hydrolysis of urea at 96°C for 96 h in the triangular flask. The pH value was the critical factor affecting the crystal growth process after fixing the Ca/P ratio condition (Kang et al. 2008).

The phase transformation using solid materials as precursors was newly developed to synthesize inorganic crystals with special morphologies. After soaking the flat-plate shapes of brushite ($CaHPO_4 \cdot 2H_2O$) in Dulbecco's modified eagle medium (DMEM) solutions at 36.5°C, the OCP with similar shapes were obtained (Mandel and Tas 2010). Nanoscale needles, fibers, and sheets of HAp were selectively hydrothermal prepared through the hydrolysis of a solid precursor crystal of monetite ($CaHPO_4$, DCP) in an alkali solution by simply varying the pH, ion concentrations, and phosphate ions. Long HAp fibers were observed under a relatively mild basic condition at pH 9–10. The fibrous morphology evolved from the nanoneedles produced by the solid–solid transformation with the elongation of the c-axis through a dissolution–precipitation route. Flaky HAp nanosheets consisting of a parallel assembly of nanoneedles were observed with an excess amount of phosphate ions under mild basic conditions. The presence of phosphate ions suppressed the solid–solid transformation and promoted the formation of a 2D structure through dissolution–precipitation process. Moreover, the oriented array of bundled nanoneedles of HAp elongated in the c-axis was obtained under a highly basic condition at pH 11–13. The ordered architecture originated from the spatially periodic nucleation of HAp seeds on the DCP surface was through topotactic solid–solid transformation (Ito et al. 2008). Furthermore, the functional ions could be incorporated into the final products via addition of the ions into the solutions (Pieters et al. 2010).

The α-triclcium phosphate [α-$Ca_3(PO_4)_2$, α-TCP] and calcium silicate with good dissolution ability provide other kinds of precursors to modulate the morphology of HAp crystals. After hydrothermal treatment of the α-TCP powders, granules, and scaffolds in aqueous, the precursors usually transformed to whisker-like HAp crystals with nano- or microstructured morphologies (Ioku et al. 2006; Wakae et al. 2008; Liu et al. 2011). Through simply adjusting the reaction temperature and the concentration of Ca^{2+} ions under hydrothermal treatment of the α-TCP powders in aqueous without using any surfactants or additives, the well-developed HAp crystals with different structures and morphologies (chrysanthemum-like HAp microflowers, enamel-like HAp microparticles, rectangle-shaped HAp microplates, and

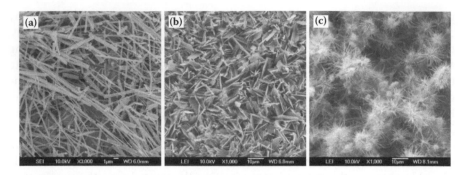

FIGURE 6.2
FESEM images of HAp powders with different morphologies: (a) rodlike, (b) platelike, (c) chrysanthemum-like.

HAp microrods) could be obtained (Figure 6.2) (Liu et al. 2011). Recently, we developed a novel process to synthesize element-substituted HAp with controllable morphologies and chemical compositions using calcium silicate as hard-template precursors in the absence of any surfactants and additives (Lin, Chang, Liu, et al. 2011). When hydrothermal treatment of the amorphous calcium silicate hydrate (CSH) precursor in Na_3PO_4 solution occurs, HAp nanoparticles with diameters about 90 nm were obtained (Figure 6.3a). In contrast, when the crystalline calcium silicate (CS) powders were used as the precursor and the Na_3PO_4 solution was used as the phosphorus source, the obtained HAp powders consisted of nanowires with lengths up to 2 μm and diameters about 100 nm (Figure 6.3b). Figure 6.3c illuminates that the obtained HAp powders via hydrothermal treatment with the crystalline CS powders in NaH_2PO_4 solutions had a smooth surface and ultralong sheetlike shape with thickness about 100 nm, widths 1~5 μm, and lengths up to 20 μm. The model of $Ca_9(PO_4)_6$ clusters (Posner clusters) with positive charge could be

FIGURE 6.3
FESEM images of HAp powders with different morphologies: (a) nanoparticles (CSH as precursor, Na_3PO_4 as solution); (b) nanowires (CS as precursor, Na_3PO_4 as solution), (c) nanosheets (CS as precursor, NaH_2PO_4 as solution).

applied to illuminate the effect of the different phosphorus sources on HAp morphology development (Posner and Betts 1975). The $Ca_9(PO_4)_6$ clusters are considered the growth unit of HAp crystals. The Ca^{2+} ions gradually dissolved from the CS precursors into the phosphate solution to form $Ca_9(PO_4)_6$ clusters. It is well known that hexagonal HAp crystal has two types of crystal surfaces with different charges, positive on a- and b-surfaces and negative on c-surfaces (Kawasaki 1991). Usually, hexagonal HAp crystals, which grow along the c-axis, are easily obtained because of a strong bond site for the $Ca_9(PO_4)_6$ cluster in the (0 0 0 1) direction but not in the (1 0 1 0) direction. The crystal growth is thus easier in the (0 0 0 1) direction than in the (1 0 1 0) direction. In the Na_3PO_4 solution, a large quantity of PO_4^{3-} ions was ionized into the solution. Therefore, there would be enough PO_4^{3-} ions to form Posner clusters with Ca^{2+} ions dissolved from CS particles. The released Ca^{2+} ions were rapidly consumed, which resulted in quite a few Ca^{2+} ions remaining near the surfaces of the CS particles. According to the Cluster growth model for HAp, Posner clusters would attach to c-surfaces preferentially and the direction along the c-axis quickly developed. Ultimately, the HAp nanowires were obtained. However, in the NaH_2PO_4 solution, the hydrolyzation of NaH_2PO_4 salts was greater than ionization. The major ions in the solution were H_2PO_4 and HPO_4^{2-} ions, and there were only a small amount of PO_4^{3-} ions formed from the secondary ionization. Therefore, a large amount of released Ca^{2+} ions were attached to the c-surfaces with negative charges, which resulted in fewer Posner clusters incorporated onto the c-surfaces. The growth of c-surfaces was limited, whereas the growth of a, b-surfaces was enhanced, leading to the aggregation of a, b planes. Ultimately, the HAp nanosheets were obtained. However, the morphology of the synthesized HAp was nanoparticles after hydrothermal treatment of the CSH powders in Na_3PO_4 solution. This might be attributed to the crystal structure of CSH itself. It is well known that the CSH is a poorly ordered phase with layered structures. The layer consists of a central Ca-O part sandwiched between parallel silicate chains (Taylor 1986). With hydrothermal treatment in phosphate solution, the HAp crystal formed accompanied with the split of the sandwiched layers and silicate chains of the CSH into the short fragmentations. In this situation, HAp crystals were formed on these fragmentations, resulting in the particle-like products. At the same time, through regulating the chemical compositions of the precursors and the reaction ratio of the precursor/solution, the HAp crystals substituted by different kinds and amount of elements (such as Si, Na, Mg, and Sr) could be easily obtained.

6.2.2 Fabrication and Morphology Control of the Nanostructured Bioceramics with Novel 3D Architectures

The three-dimensional (3D) architectured biomaterials with nanostructures have been attracting intense interests due to their high specific surface area and novel 3D hierarchical architectures, which make it possible to incorporate higher dosages of drugs, protein, or DNA molecules into the hierarchical structures and release them at a controlled rate (Zhang, Cheng, et al. 2009; Zhang, Yang, et al. 2009; Lin, Chang, and Zhu 2011; Lin, Zhou, et al. 2011; Wu et al. 2011). Furthermore, the 3D architectured materials can be used as injectable bone regeneration biomaterials and cell- or drug-loaded implants, which are superior to particles. Therefore, the fabrication of nanostructured biomaterials with novel 3D architectures is of great importance for enhancing performances and expanding applications of them in biomedical fields. The traditional strategy to synthesize the 3D architectured inorganic biomaterials is focused on the surfactant assistant assembled approach (He et al. 2007; Ma and Zhu 2009; Wang et al. 2009; Zhang, Chang, et al. 2009; Zhang, Yang, et al. 2009; Cheng et al. 2010; He et al. 2010). As a recently developed concept, the self-assembly technologies using nanoparticles, nanorods, nanobelts, and nanosheets as building blocks have been shown to be an efficient "bottom-up" route to fabricate functional materials with different morphologies and architectures (Corma et al. 2004; Lu et al. 2004). The surfactant assistant self-assembly method has been considered the most effective method to control the morphology and architectures of the functional materials. The surfactant as a capping reagent plays an important role in the process of self-assembly or oriented aggregation, facilitating the formation of special morphologies and hierarchical superstructures to reduce the surface energy and thus the total system energy, through the dipole–dipole interaction or van der Waals forces. Moreover, it was revealed that the very short ligand greatly decreased the distance between the primary particles and thereby enhanced the dipole interaction between them (Narayanaswamy et al. 2006; Lin, Chang, et al. 2009). On the other hand, the van der Waals attractions between the surfactant and the nanoparticles could also be responsible for the morphology evolution.

Using EDTA, Na_2EDTA, and CA, for example, as the chelate and template-directed reagents, and usually using urea as the homogeneous precipitate reagent, the apatite materials with dandelion-like (Lak et al. 2008), porous (He and Huang 2007, 2009), dumbbell-like (Chen et al. 2006), flowerlike (Chen et al. 2006), and hollow nanostructures constructured by rods, whiskers, and nanosheets were successfully synthesized via the hydrothermal process. The studies showed that the template and chelate type, pH values, and temperatures play important roles in modulating the architectures of the products. He and Huang (2009) found that the template Na_2EDTA altered the natural growth habit of carbonated HAp (CHAp) and induced the first dendritic growth of CHAp nanoflakes on the surface of OCP, and the organic conditioner glycerol controlled the second dendritic growth of CHAp nanoflakes on the bigger ones. Higher concentrations of glycerol resulted in a higher surface fractal dimension, smaller pore diameter, and greater specific surface area. Chen et al. (2009) found that the morphology of fluorapatite (FHAp) nanocrystal was highly dependent on the pH value and nature of chelating reagent. In the case of CA, two main morphologies of FHAp nanocrystals were observed as the pH value varied from 3.6 to 10.0. At low pH 3.6, the FHAp product was composed of uniform 3D nanorod aggregates, which were composed of numerous well-aligned nanorods. With the increase of the pH to 5.2, the FHAp was composed of dumbbell-like crystals as well as some particles with irregular shape. With further increase of the pH, the aggregates disappeared and ultimately resulted in morphology of short nanorods with a mean length of about 50 nm under pH 10.0. One-dimensional FHAp nanocrystals with hexagonal structure were always produced in the presence of Na_2EDTA under different pH values. When the two chelating reagents of CA and Na_2EDTA were simultaneously used, the flowerlike crystals with shuttle-shaped petals were obtained at pH 3.6. With the increase of the pH to 5.2, the FHAp sample was made up of 1D hexagonal nanorods with about 2 μm in length and 300 nm in diameter. The well-aligned nanorods of FHAp were obtained at pH 6.1. With further increase of the pH to 10.1, the sample is composed of a wealth of branchlike nanorods. It implied that the combination of these two chelating reagents induced not only morphology change but also structural transformation. The possible mechanism for the control of FHAp crystals with various morphologies might be attributed to the preferential crystal growth habit along the (0 0 0 1) direction for FHAp, possible anisotropic growth along crystallographically reactive directions under different pH and in the presence of chelating reagent, the molecule structures, ionization, and surface adsorption of the chelating reagents at different pH conditions. The kinetic analysis of HAp crystal growth also revealed that the template depressed the interfacial potential energy (E), then enhanced the roughness on the surface of crystal nucleus and directed HAp crystal to selectively grow along the (0 0 0 1) direction, and consequently governed the aperture of porous HAp microspheres (He and Huang 2007).

The HAp microtubes were prepared using $CaCl_2$ and NaH_2PO_4 in mixed solvents of water–N, N-dimethylformamide (DMF) at ratio of 1:2 by a solvothermal method at 160°C for 24 h. The concentration of DMF played an important role on the morphologies of the product (Ma et al. 2008). The flowerlike nanostructured HAp hollow spheres (NHHS) assembled with nanosheets with a hierarchical morphology were fabricated via a rapid microwave-assisted hydrothermal route. The presence and concentration of block copolymer poly(lactide)-blockpoly(ethylene glycol) (PLA-PEG) were important parameters for the self-assembly of the hollow structure (Wang et al. 2010). The polyelectrolytes have been widely used as templates to synthesize and modulate the morphologies of the advanced inorganic materials. Wang et al. (2009) used negatively charged poly(styrene sulfonate) (PSS) to modify the morphology and particle size of HAp by simply adjusting the PSS concentration. Within the PSS concentration range of 0 to 9.6 wt%, the HAp crystallites grew from ribbons to microspheres, and the building units of various microspheres changed from nanofibers to nanorods or nanoplates. Along with that, the microspheres became smaller and more compact at higher PSS concentrations. The adsorption of PSS onto certain crystal faces as well as the complex effect of PSS with Ca^{2+} could be considered as the controlling factors that determined the influence of PSS on the crystal growth mode. The flexible polymer with functional groups, such as polyethylene glycol (PEG), can be also used as the cosurfactant or cotemplate to modulate the morphology and architecture of the products (Salarian et al. 2009). Spoerke et al. (2009) designed a scaffolding framework assembled from the peptide amphiphile (PA) nanofiber, which was mimetic to the natural collagen. The research shown that enzymatically mediated harvesting of phosphate ions combined with nanofiber surface nucleation could lead to a spatially selective and biomimetic mineralization in a 3D environment. Their study suggests that both spatial and temporal elements are necessary to achieve biomimetic mineralization in synthetic materials.

The solution mediated deposition method was also used to fabricate the inorganic materials. Schmidt et al. (2004; Schmidt and Ostafin 2002) coated apatite on lipsome micelles to form nanoshells, and the thickness of shells could be well controlled by adjusting the addition time of calcium and phosphate salt. Similarly, Tjandra et al. (2006) directly precipitated apatite on block copolymer templates, and hollow apatite nanospheres were obtained after calcination. Fujii et al. (2009) fabricated HAp-coated micrometer-sized poly(L-lactic acid) (PLLA) microspheres via a "Pickering-type" emulsion route in the absence of any molecular surfactants. In their strategy, the stable oil-in-water emulsions were prepared using 40 nm HAp nanoparticles as a particulate emulsifier and a dichloromethane (CH_2Cl_2) solution of PLLA as an oil phase, in which the interaction between carbonyl/carboxylic acid groups of PLLA and the HAp nanoparticles at the CH_2Cl_2–water interface played a crucial role to prepare the stable Pickering-type emulsion. After evaporation of CH_2Cl_2 from the emulsion, the HAp nanoparticle-coated PLLA microspheres were fabricated.

Recently, morphology control of bioinorganic materials through biomimetic approaches is one of the most interesting issues in the material fabrication and application fields, in which the inorganic materials with complex morphologies can be fabricated. Fowler et al. (2005) synthesized enamel-like HAp with bundle structures directly from a solution containing AOT, water, and oil. The bundles were only 750 nm to 1 mm in length and 250 to 350 nm in width. In biological bone, the composite structure of nanosized HAp reinforced collagen offers a route to manufacture materials with high strength and high toughness. The micrometer spherical HAp crystals with the diameter of 1 to 3 μm were synthesized via a biomimetic process using 1.5% β-cyclodextrin (β-CD) as a template (Xiao et al. 2009). The interaction between Ca^{2+} ions and function groups –OH in β-CD reduced the drifting velocity and the concentration of ions, which resulted in the decrease of electrical conductivity. This interaction between Ca^{2+} and β-CD is possibly one of the influence factors on the nucleation and growth of the spherical HAp crystal. In addition, β-CD is a cyclic oligosaccharide consisting of seven α-(1-4)-linked D-glucose units. This special molecular structure provides a molecule shaped like a segment of a hollow cone with an exterior hydrophilic

FIGURE 6.4
FESEM images of the flowerlike HAp powders.

surface, and the β-CD molecules will cross-link with each other to form the spherelike structure through the hydroxyls.

In most of these cases, large amounts of surfactant and soft templates, and even unhealthy templates or organic solvents were widely used, which were hazardous to health and environment. Up to now, self-assembly of nanounits into the 3D architectures in the absence of any surfactants, template support, or structure-directing reagents are still a major challenge. In the absence of any surfactants, template support, and structure-directing reagents, uniform 3D structured carbonated apatite flowers (Figure 6.4) with exclusively nanosheet-assembled network morphology were synthesized by Lin et al. via a low-temperature hydrothermal process, using $Ca(NO_3)_2$ and $NH_4H_2PO_4$ as Ca and P sources, and urea as the homogeneous precipitation reagent. The uniformity in size distribution and shape of the apatite flowers was probably attributed to the homogeneous precipitation effects, and the high crystallization of the products was attributed to the hydrothermal treatment. A possible self-assembled mechanism was preliminarily proposed for the formation of the novel 3D architectures. The process of the morphology evolution for HAp flowers was proposed as the nucleation–dissolution–recrystallization–self-assembly growth from nanosheets to a flowerlike morphology process (Figure 6.5) based the basis of time-dependent experiments and the morphology observations.

Furthermore, we also synthesized the biomimetic HAp porous microspheres with cosubstituted essential trace elements (Na, Mg, K, F, Cl, and CO_3^{2-}) of natural bone via a low-temperature hydrothermal method (Figure 6.6), in absence of any surfactants, organic solvent, or template-directing reagents. Comprehensive studies have confirmed that these substitutions play critical roles in biological performances of the materials in comparison to stoichiometric HAp. In our synthesis process, the $Ca(NO_3)_2$ and $NH_4H_2PO_4$ were used as Ca and P sources, respectively; $NaNO_3$,

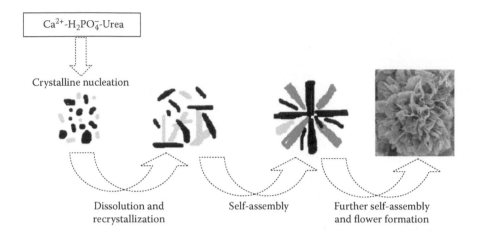

FIGURE 6.5
Schematic illustration of the formation and morphology evolution of 3D structured HAp flowers in the whole synthetic process.

$Mg(NO_3)_2$, KNO_3, NH_4Cl, and NH_4F were used as substituted ion sources; and urea was applied as the homogeneous precipitation reagent and CO_3^{2-} source. The characterization results showed that the obtained biomimetic HAp porous microspheres were self-assembled by 2D single crystalline nanosheets. The novel 3D architectures resulted in favorable drug loading and release properties, and the cosubstituted essential trace elements decreased the crystallinity of the products and enhanced the degradability of the porous microspheres in comparison with the traditional pure HAp

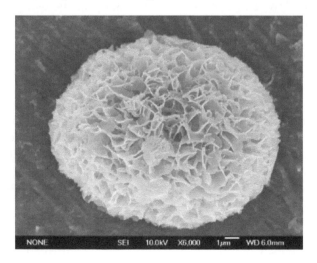

FIGURE 6.6
FESEM image of the synthetic HAp porous microspheres with cosubstituted essential trace elements (Na, Mg, K, F, Cl, and CO_3^{2-}) of natural bone.

materials (Lin, Zhou, et al. 2011). Wang et al. (2011) reported a template-free microwave-assisted hydrothermal method for the preparation of HAp hollow microspheres constructed by the self-assembly of nanosheets using $Ca(CH_3COO)_2$, Na_2HPO4, NaH_2PO_4, and sodium citrate in aqueous solution. During the microwave-hydrothermal process, HAp nanosheets were formed first, and then these HAp nanosheets were bound together toward microaggregates through the interaction between $-COO$ groups of Na_3Cit molecules and Ca^{2+} ions of the HAp nanosheets, thus an assembling process occurred.

The hard templates were also used to fabricate HAp hollow nanostructures. Recently, we developed a facile strategy to delicately control the morphologies of HAp materials from simple 0D morphologies to complicated 3D architectures using hard precursors with similar structures, which provided a new platform for HAp materials to be efficiently synthesized and manipulated (Lin, Chang, and Zhu 2011). The HAp nanoparticles, nanowires, and hollow nanostructured microspheres (Figure 6.7) were facilely synthesized via hydrothermal treatment of the similar structured precursors of calcium carbonate ($CaCO_3$) nanoparticles, xonotlite $[Ca_6(Si_6O_{17})(OH)_2]$ nanowires, and hollow $CaCO_3$ microspheres in Na_3PO_4 solutions, respectively. In addition, their sizes could be easily regulated by changing the template conditions. Li and Hashida (2006) further fabricated HAp-whisker ceramics with a hydrothermal hot-pressing (HHP) method using α-TCP as the precursors and in water or ammonia water condition. When the additive was water, the obtained HAp grains showed whisker-like and platelike features. When the additive was ammonia water, the resultants consisted of Hap whiskers only. The results show that treating α-TCP with addition of ammonia water by the HHP method is a useful method for *in situ* fabrication of HAp whisker ceramics.

The alloys have been widely used as the load-bearing implants in clinical application fields. However, the bioinert of the alloys limits their wider applications. The Ca-P–based bioceramic coating on alloys is one of the most popular methods to improve the bioactivity of the alloys. Several methods have been explored to deposit Ca-P coatings in order to enhance implant

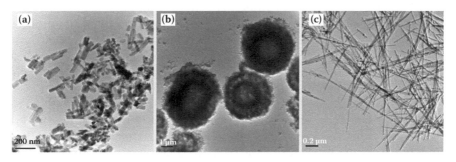

FIGURE 6.7
Morphologies of the HA nanoparticles, nanowires, and nanostructured hollow HAp microspheres transformed from $CaCO_3$ nanoparticles, $Ca_6(Si_6O_{17})(OH)_2$ nanowires, and hollow $CaCO_3$ microsphere precursor, respectively.

fixation. SBF soaking, hydrothermal homogeneous precipitation, and electrodeposition have been widely used due to (1) the methods enable formation of highly crystalline deposits with low solubility in body fluids and low residual stresses; (2) the ability to coat porous, geometrically complex, or non-line-of-sight surfaces; (3) the ability to control the thickness, composition, and microstructure of the deposits; (4) the possible improvement of the substrate and coating bond strength and (5) the availability and low cost of equipment. The compositions of the precursor solutions, pH values, and additives were found to significantly affect the Ca-P crystal phases, surface morphologies, crystal sizes, and chemical compositions of the Ca-P deposits in the electrocrystallization process (Eliaz and Sridhar 2008; Haders et al. 2008; Stenport et al. 2008).

Electrospinning is good technology to fabricate ultrafine and ultra long fibers with diameters from nano- to micrometers under high electrostatic force. The main features of nonwoven fabrics composed of the fibers include large surface area, high percentage of voids, and small pore size. These features are highly desirable in scaffolds for tissue engineering and in substrates for functional cell cultures such as in three-dimensional model systems. Such scaffolds and substrates have been developed from a variety of materials, including polymers, inorganics, and their composites (Ramanan and Venkatesh 2004; Wu et al. 2004; Schnell et al. 2007; Chew et al. 2008; Yamaguchi et al. 2008; Franco et al. 2012; Xie et al. 2012; Zou et al. 2012). Recently, the bioactive silicate fiber scaffolds were fabricated with a combination of electrospinning technology and the sol-gel process (Yamaguchi et al. 2008; Franco et al. 2012). The HAp fibers with porous surface and uniform fiber diameter of about 140 μm were fabricated by the electrospinning sols of the 2-butanol solution of phosphorous pentoxide and calcium acetate solution in distilled water, and lactic acid was added as a spinning aid. Yamaguchi et al. (2008) prepared silicate electrospun nanofibers via the sol-gel process. Xie et al. (2012) prepared the submicron bioactive glass tubes using sol-gel and coaxial electrospinning techniques for applications in bone tissue engineering. The heavy mineral oil and gel solution were delivered by two independent syringe pumps during the coaxial electrospinning process. Subsequently, submicron bioactive glass tubes were obtained by removal of poly(vinyl pyrrolidone) and heavy mineral oil via calcination at 600°C for 5 h. The tubular structures possess high surface area, high protein-loading capacity, and delayed release properties.

We prepared the nanocrystalline HAp assembled hollow fibers (NHAHF) in the membrane form by combining the electrospinning technique and the hydrothermal method. First, the electrospun bioactive glass fibers (BGF) were prepared via mixtures of the bioactive glass precursor (molar ratio of $SiO_2:CaO:P_2O_5 = 70:25:5$) and poly(vinyl butyral) (PVB). After calcining at 600°C for 4 h in air, the polymer PVB was removed and the resulting bioactive glass nanofiber nonwoven membrane was obtained and then was used as a self-sacrificial template. After hydrothermal treatment the bioactive

FIGURE 6.8
SEM images of the (a) electrospun BGF and (b) the obtained NHAHF after hydrothermal treatment at 150°C for 2 h.

glass nanofiber nonwoven membrane in $(NH_4)_2HPO_4$ aqueous solution at 150°C, the hierarchically structured nanocrystalline HAp assembled hollow fibers were fabricated (Figure 6.8). Analogous to conventional hard template, the bioactive glass nanofibers in our method cannot only play the role of a structure-directing scaffold and a precursor for the shell but also automatically dissolve during the shell-forming process. The progress is therefore more efficient and the resulting hollow structure is continuous, uniform in a larger scale, and controllable by changing the reaction time (Wu et al. 2011).

6.2.3 Morphology Control of the Bioceramics with Oriented Structures Similar to Bone and Tooth

The natural human bone and tooth possess a highly organized array of HAp crystallites on the nanoscale level and bundles of aligned crystallites woven into intricate architectures on the microscale level (Habelitz et al. 2001; LeGeros 2008). Such a unique hierarchical structure plays a critical role in determining the excellent mechanical properties and biological properties of them. It is also a great challenge to develop facile methods to fabricate HAp materials with oriented hierarchical structure similar to bone and tooth (Chen et al. 2006; Yang et al. 2011). It is clear that the processes of HAp crystal nucleation and growth in the extracellular matrix are highly controlled and directed by proteins. Therefore, the most popular strategy to synthesize bonelike and enamel-like HAp is based on the protein-directing process (Moradian-Oldak 2001). In enamel, processing of the matrix is concomitant with the growth and maturation of enamel crystallites, and perhaps the step-wise processing of amelogenin is one of the key factors to control some processes of crystal growth. Amelogenin sequences at both carboxy- and amino terminal regions across species are highly homologous suggesting that these regions play specific functional roles during matrix mediated enamel biomineralization (Toyasawa et al. 1998; Moradian-Oldak 2001). Wen et al. (2000)

found that after the addition of 1%~2% amelogenins to 10% gelatin gel the OCP crystals became remarkably longer. The average aspect ratio (length/ width) of OCP increased remarkably with the increase of the concentration of amelogenins, which suggested that the elongation effect was dose dependent by amelogenin. Iijima et al. (2001) also found that the ratios of the length-to-width and the thickness-to-width of the OCP crystals were increased in the presence of amelogenins isolated from developing bovine enamel, as well as recombinant amelogenins rM179 and rM166. These studies indicate that amelogenin interacts more strongly with the (0 1 0) face, which inhibit the growth of the crystals in the b-axial direction (the width). Analysis of the particle size distribution of apatite crystal aggregates grown in the presence of different amelogenins indicated that the full-length amelogenin caused aggregation of apatite crystals more effectively than the C-terminally cleaved amelogenin and other commercially available macromolecules (Moradian-Oldak et al. 1998). The enamel-liked HAp aggregations were also reported by using bovine amelogenin. The results also reveal that, as a well-known effective modifier during *in vivo* tooth enamel formation, the amelogenin can dramatically accelerate the kinetics of nanoassembly (Wang et al. 2007).

Tao et al. (2007) suggested a new model of "bricks and mortar" concerning the biological aggregation of apatite nanoparticles. An inorganic phase, amorphous calcium phosphate (ACP) acts as "mortar" to cement the crystallized "bricks" of nano-HAp. Meanwhile, biological molecules control the nanoconstruction. By using HAp nanospheres as the building blocks, highly ordered enamel-like and bonelike apatite were hierarchically constructed in the presence of glycine (Gly) and glutamate (Glu), respectively. It is interesting that during the evolution of biological apatite, the amorphous mortar can be eventually turned into the brick phase by phase transformation to ensure the integrity of biominerals. Cai et al. (2009) simulated the assembly of nanosized HAp crystals using the silk sericin (SS) as template. The SS concentration and mineralization time played important roles on crystal size, morphology, and formation process (such as nucleation, growth, aggregation, especially assembly dynamics, etc.) of the products. When the concentration of SS was kept at 1% (w/v) in the reaction system, the HAp crystals of 300 to 500 nm in length and 50 to 80 nm in diameter were assembled along the c-axis and smaller crystals of about 20 nm were obtained. The size and the assembled microstructures are similar to the natural enamel crystals. The original self-assembled spherical-shape and subsequent conformational transition of SS may be two major reasons to modulate the enamel prismlike structure formation.

The enamel-like HAp crystals were synthesized by modifying the synthetic HAp nanorods with a surfactant of AOT, which allowed the nanorods to self-assemble into an enamel prismlike structure at the water–air interface (Tornblom and Henriksson 1997). Ye et al. (2008) achieved highly oriented organization of HAp nanorods through a simple reflux method using mixtures of triblock copolymer pluronic P123 and tween-60 as the mediated reagents.

Li et al. (2007) showed that F$^-$ ion played a marked effect on the composition and morphology of deposited HAp crystals. In the absence of F$^-$ ions, spherical morphology HAp containing CO_3^{2-} was formed on the sulfonic-terminated self-assembled monolayer (SAM). When F$^-$ was added, HAp crystals containing both CO_3^{2-} and F$^-$ were formed on the SAM. Needle-shaped crystals of high aspect ratio and 1 to 2 μm in length grew elongated along the c-axial direction in bundles, mimicking HAp crystals in tooth enamel. The formation of enamel-like HAp can be attributed to the substitute of F$^-$ for OH$^-$ by disturbing the normal progress of HAp formation on the SAM.

Yuan et al. (2008), using porous anodic aluminum oxide (AAO) as a template, fabricated a highly ordered array of HAp nanotubes with uniform length and diameter using the sol-gel autocombustion method. A potential mechanism of "an autocombustion from dried gel to nanoparticles and a subsequent *in situ* reaction from nanoparticles to nanotubes" was proposed based on the electron diffraction, x-ray diffraction (XRD), and x-ray photoelectron spectroscopy survey.

Other than the polymers and proteins, the small molecules with special structures and functional groups can also be used as the template to direct the crystal growth and their aggregations. Compared with the traditional proteins, the small molecules process has more robust physicochemical properties, such as tolerance of temperature and resistance to acid or base, which would allow it to be better applied in wider and more severe reaction conditions. Recently, Liu et al. (2012) synthesized the highly oriented HAp arrays (Figure 6.9b) with the aid of small quercetin ($C_{15}H_{10}O_7$, QUE; Figure 6.9a) molecule via hydrothermal treatment of α-TCP precursors in a QUE-containing ethanol–distilled water solution. Furthermore, the morphologies of HAp nanocrystals varied from wide-angle branching to small-angle branching, and finally parallel-packed arrays could be well modulated by simply regulating the concentration of QUE. The HAp morphology regulating might be attributed to the physicochemical properties of QUE. The

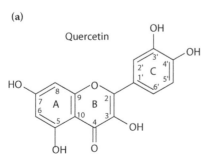

FIGURE 6.9
(a) Molecular structure of QUE and (b) FESEM image of samples obtained under 180°C for 24 h with QUE concentration of 50.00 mg/20 mL, top-right insert: higher resolution image.

QUE is a naturally occurring flavonoid (also belongs to polyphenols) in our daily food, which has a planar structure and tends to form highly oriented nanostructures in suspension (Sahoo et al. 2011). The molecular size of QUE is in a similar dimension to the unit cell formed by Ca^{2+} ions on the (1 1 0) faces (a, b plane) of HAp (Posner and Betts 1975; Dey et al. 2010). The negative charge of the QUE molecule plane makes it attracted to the positively charged a, b plane of HAp. Therefore, the specific molecular complementarity between QUE and HAp associated with the negative charge property of QUE played very important roles in the orientation control of HAp crystal growth, for they enabled QUE to carry and release Ca^{2+} ions and attach to the HAp surface under the hydrothermal reaction environment with different QUE concentrations.

6.3 Conclusion and Perspective

Bioactive bioceramics are widely used in hard tissue repair and regeneration, and also in drug and gene delivery systems. The performance of bioceramics in their applications depends greatly on its crystal morphologies, particle sizes and size distribution, aggregates, and 3D architectures. Various strategies have been developed to control the morphology and 3D architectures of the bioceramic materials, and the mechanisms behind them might be very different. In this chapter, we summarize the preparation and mechanism of bioceramic materials with controllable morphology and crystal growth. However, there are still many open questions and challenges within this field that need to be further investigated in detail. The real mechanisms for the controlling of the morphology and crystal growth need to be comprehensively researched and confirmed. It is still difficult to synthesize the uniform bioceramic particles with monodispersion and narrow-size distribution in large scale, which is very important in drug and gene delivery system applications. The fabrication of bonelike and enamel-like bioceramics with highly oriented architectures also remains a great challenge. Up to now, few works have been reported on the fabrication of large size (such as in centimeter size) bioceramics and scaffolds with nano- and microstructured surfaces. It could be concluded that the self-assembly and biomineralization methods are the direction of the future in the morphology control for bioceramics. Furthermore, more and arduous works in the study of the relationships and interactions between the morphologies, grain sizes, crystal aggregations, architectures, and the cell and tissue biological responses remain to be undertaken.

References

Arce, H., Montero, M. L., Saenz, A., et al. 2004. Effect of pH and temperature on the formation of hydroxyapatite at low temperatures by decomposition of a Ca–EDTA complex. *Polyhedron* 23: 1897–901.

Bang, J., and Suslick, K. 2010. Applications of ultrasound to the synthesis of nanostructured materials. *Adv. Mater.* 22: 1039–59.

Bocciarelli, D. 1970. Morphology of crystallites in bone. *Calcif. Tissue Int.* 5: 261–9.

Bose, S., and Saha, S. K. 2003. Synthesis and characterization of hydroxyapatite nanopowders by emulsion technique. *Chem. Mater.* 15: 4464–9.

Bouyer, E., Gitzhofer, F., and Boulos, M. I. 2000. Morphological study of hydroxyapatite nanocrystal suspension. *J. Mater. Sci. Mater. Med.* 11: 523–31.

Burke, E. M., Guo, Y., Colon, L., et al. 2000. Influence of polyaspartic acid and phosphophoryn on octacalcium phosphate growth kinetics. *Colloids Surf. B* 17: 49–57.

Byrappa, K. 2001. *Handbook of hydrothermal technology.* Noyes Publications, Norwich, New York.

Byrappa, K., and Adschiri, T. 2007. Hydrothermal technology for nanotechnology. *Prog. Cryst. Growth Ch. Mater.* 53: 117–66.

Cai, Y., Jin, J., Mei, D., et al. 2009. Effect of silk sericin on assembly of hydroxyapatite nanocrystals into enamel prism-like structure. *J. Mater. Chem.* 19: 5751–8.

Cai, Y., Liu, Y., Yan, W., et al. 2007. Role of hydroxyapatite nanoparticle size in bone cell proliferation. *J. Mater. Chem.* 17: 3780–7.

Chander, S., and Fuerstenau, D. W. 1984. Solubility and interfacial properties of hydroxyapatite: A review. In D. N. Missra (ed.), *Adsorption on and surface chemistry of hydroxyapatite*, 29–49. Plenum Press, New York.

Chen, D., Gao, L., and Zhang, P. 2003. Synthesis of nickel sulfide via hydrothermal microemulsion process: Nanosheet to nanoneedle. *Chem. Lett.* 32: 996–7.

Chen, H., Tang, Z., Liu, J., et al. 2006. Acellular synthesis of a human enamel-like microstructure. *Adv. Mater.* 18: 1846–51.

Chen, M., Jiang, D., Li, D., et al. 2009. Controllable synthesis of fluorapatite nanocrystals with various morphologies: Effects of pH value and chelating reagent. *J. Alloy. Comp.* 485: 396–401.

Cheng, X., Huang, Z., Li, J., et al. 2010. Self-assembled growth and pore size control of the bubble-template porous carbonated hydroxyapatite microsphere. *Cryst. Growth Des.* 10: 1180–8.

Chew, S. Y., Mi, R., Hoke, A., et al. 2008. The effect of the alignment of electrospun fibrous scaffolds on Schwann cell maturation. *Biomaterials* 29: 653–61.

Corma, A., Rey, F., Rius, J., et al. 2004. Supramolecular self-assembled molecules as organic directing agent for synthesis of zeolites. *Nature* 431: 287–90.

Dey, A., Bomans, P. H. H., Muller, F. A., et al. 2010. The role of prenucleation clusters in surface-induced calcium phosphate crystallization. *Nat. Mater.* 9: 1010–4.

Diegmueller, J. J., Cheng, X., and Akkus, O. 2009. Modulation of hydroxyapatite nanocrystal size and shape by polyelectrolytic peptides. *Crystal Growth Des.* 9: 5220–6.

Dorozhkin, S. 2007. Calcium orthophosphates. *J. Mater. Sci.* 42: 1061–95.

Eliaz, N., and Sridhar, T. M. 2008. Electrocrystallization of hydroxyapatite and its dependence on solution conditions. *Cryst. Growth Design.* 8: 3965–77.

Fowler, C., Li, M., Mann, S., et al. 2005. Influence of surfactant assembly on the formation of calcium phosphate materials—A model for dental enamel formation. *J. Mater. Chem.* 15: 3317–25.

Franco, P. Q., João, C. F. C., Silva, J. C., et al. 2012. Electrospun hydroxyapatite fibers from a simple sol-gel system. *Mater. Lett.* 67: 233–6.

Fujii, S., Okada, M., Sawa, H., et al. 2009. Hydroxyapatite nanoparticles as particulate emulsifier: Fabrication of hydroxyapatite-coated biodegradable microspheres. *Langmuir* 25: 9759–66.

Habelitz, S., Marshall, S. J., Marshall Jr., G. W., et al. 2001. Mechanical properties of human dental enamel on the nanometre scale. *Arch. Oral Biol.* 46: 173–83.

Haders, D. J., Burukhin, A., Zlotnikov, E., et al. 2008. TEP/EDTA doubly regulated hydrothermal crystallization of hydroxyapatite films on metal substrates. *Chem. Mater.* 20: 7177–87.

He, Q., and Huang, Z. 2007. Controlled growth and kinetics of porous hydroxyapatite spheres by a template-directed method. *J. Cryst. Growth* 300: 460–6.

He, Q., and Huang, Z. 2009. Controlled synthesis and morphological evolution of dendritic porous microspheres of calcium phosphates. *J. Porous Mater.* 16: 683–9.

He, Q., Huang, Z., Liu, Y., et al. 2007. Template-directed one-step synthesis of flower-like porous carbonated hydroxyapatite spheres. *Mater. Lett.* 61: 141–3.

He, W. H., Tao, J. H., Pan, H. H., et al. 2010. A size-controlled synthesis of hollow apatite nanospheres at water-oil interfaces. *Chem. Lett.* 39: 674–5.

Hench, J. 1991. Bioceramics: From concept to clinic. *J. Am. Ceram. Soc.* 74: 1487–510.

Hench, L., and Polak, J. 2002. Third-generation biomedical materials. *Science* 295: 1014–7.

Huang, S., Zhu, J., and Zhou, K. 2012. Effects of Eu^{3+} ions on the morphology and luminescence properties of hydroxyapatite nanoparticles synthesized by one-step hydrothermal method. *Mater. Res. Bull.* 47: 24–28.

Iijima, M., Moriwaki, Y., Takagi, T., et al. 2001. Effects of bovine amelogenins on the crystal morphology of octacalcium phosphate in a model system of tooth enamel formation. *J. Cryst. Growth* 222: 615–26.

Iijima, M., Nelson, D. G. A., Pan, Y., et al. 1996. Fluoride analysis of apatite crystals with a central planar OCP inclusion: Concerning the role of F^- ions on apatite/OCP/apatite structure formation. *Calcif. Tissue Int.* 59: 377–84.

Ioku, K., Kawachi, G., and Sasaki, S. 2006. Hydrothermal preparation of tailored hydroxyapatite. *J. Mater. Sci.* 41: 1341–4.

Ito, H., Oaki, Y., and Imai, H. 2008. Selective synthesis of various nanoscale morphologies of hydroxyapatite via an intermediate phase. *Cryst. Growth Des.* 8: 1055–9.

Kanchana, P., and Sekar, C. 2010. Influence of sodium fluoride on the synthesis of hydroxyapatite by gel method. *J. Cryst. Growth* 312: 808–16.

Kang, Y., Yin, G., Liu, Y., et al. 2008. The precipitation of three Ca-P phase whiskers from an acid solution through hydrolysis of urea. *J. Ceram. Process. Res.* 9: 162–6.

Kawasaki, T. 1991. Hydroxyapatite as a liquid chromatography packing. *J. Chromatogr.* 544: 147–84.

Kobayashi, T., Ono, S., Hirakura, S., et al. 2012. Morphological variation of hydroxyapatite grown in aqueous solution based on simulated body fluid. *CrystEngComm* 12: 1143–9.

Koutsoukos, P. G., and Nancollas, G. H. 1981. The morphology of hydroxyapatite crystals grown in aqueous solution at 37°C. *J. Cryst. Growth* 55: 369–75.

Kuboki, Y., Takita, H., Kobayashi, D., et al. 1998. BMP-induced osteogenesis on the surface of hydroxyapatite with geometrically feasible and nonfeasible structures: Topology of osteogenesis. *J. Biomed. Mater. Res.* 39: 190–9.

Kumar, R., Prakash, K. H., Cheang, P., et al. 2004. Temperature driven morphological changes of chemically precipitated hydroxyapatite nanoparticles. *Langmuir* 20: 5196–200.

Lak, A., Mazloumi, M., Mohajerani, M., et al. 2008. Self-assembly of dandelion-like hydroxyapatite nanostructures via hydrothermal method. *J. Am. Ceram. Soc.* 91: 3292–7.

LeGeros, R. Z. 2008. Calcium phosphate-based osteoinductive materials. *Chem. Rev.* 108: 4742–53.

Li, C., Liu, S., Li, G., et al. 2010. Hydrothermal synthesis of large-sized hydroxyapatite whiskers regulated by glutamic acid in solutions with low supersaturation of precipitation. *Adv. Powder Technol.* 22: 537–43.

Li, H., Huang, W., Zhang, Y., et al. 2007. Biomimetic synthesis of enamel-like hydroxyapatite on self-assembled monolayers. *Mater. Sci. Eng. C* 27: 756–61.

Li, J. G., and Hashida, T. 2006. *In situ* formation of hydroxyapatite-whisker ceramics by hydrothermal hot-pressing method. *J. Am. Ceram. Soc.* 89: 3544–6.

Li, S. K., Guo, X., Wang, Y., et al. 2011. Rapid synthesis of flowerlike Cu_2O architectures in ionic liquids by the assistance of microwave irradiation with high photochemical activity. *Dalton Trans.* 40: 6745–6750.

Li, Y., Cai, Y., Kong, X., et al. 2008. Anisotropic growth of hydroxyapatite on the silk fibroin films. *Appl. Surf. Sci.* 255: 1681–5.

Lim, G. K., Wang, J., Ng, S. C., et al. 1997. Processing of hydroxyapatite via microemulsion and emulsion routes. *Biomaterials* 18: 1433–9.

Lin, K., Chang, J., Chen, G., et al. 2007. A simple method to synthesize single-crystalline β-wollastonite nanowires. *J. Cryst. Growth* 300: 267–71.

Lin, K., Chang, J., and Cheng, R. 2007. *In vitro* hydroxyapatite forming ability and dissolution of tobermorite nanofibers. *Acta Biomater.* 3: 271–6.

Lin, K., Chang, J., Cheng, R., et al. 2007. Hydrothermal microemulsion synthesis of stoichiometric single crystal hydroxyapatite nanorods with mono-dispersion and narrow-size distribution. *Mater. Lett.* 61: 1683–7.

Lin, K., Chang, J., Liu, X., et al. 2011. Synthesis of element-substituted hydroxyapatite with controllable morphology and chemical composition using calcium silicate as precursor. *CrystEngComm* 13: 4850–5.

Lin, K., Chang, J., and Lu, J. 2006. Synthesis of wollastonite nanowires via hydrothermal microemulsion methods. *Mater. Lett.* 60: 3007–10.

Lin, K., Chang, J., Zhu, Y., et al. 2009. A facile one-step surfactant-free and low-temperature hydrothermal method to prepare uniform 3D structured carbonated apatite flowers. *Cryst. Growth Des.* 9: 177–81.

Lin, K., Chang, J., and Zhu, Y. 2011. Facile synthesis of hydroxyapatite nanoparticles, nanowires and hollow nano-structured microspheres using similar structured hard-precursors. *Nanoscale* 3: 3052–5.

Lin, K., Chen, L., and Chang, J. 2012. Fabrication of dense hydroxyapatite nanobioceramics with enhanced mechanical properties via two-step sintering process. *Int. J. Appl. Ceram.* 9: 479–85.

Lin, K., Pan, J., Chen, Y., et al. 2009. Study the adsorption of phenol from aqueous solution on hydroxyapatite nanopowders. *J. Hazard. Mater.* 161: 231–40.

Lin, K., Zhou, Y., Zhou, Y., et al. 2011. Biomimetic hydroxyapatite porous microspheres with co-substituted essential trace elements: Surfactant-free hydrothermal synthesis, enhanced degradation and drug release. *J. Mater. Chem.* 21:16558–65.

Liu, D., Troczynski, T., Tseng, W. 2001. Water-based sol-gel synthesis of hydroxyapatite: Process development. *Biomaterials* 22: 1721–30.

Liu, X., Lin, K., and Chang, J. 2011. Modulation of hydroxyapatite crystals formed from α-tricalcium phosphate by surfactant-free hydrothermal exchange. *CrystEngComm* 13: 1959–65.

Liu, X., Lin, K., Qian, R., et al. 2012. Growth of highly oriented hydroxyapatite arrays tuned by quercetin. *Chem. Eur. J.* 18: 5519–23.

Lu, W., Gao, P., Jian, W., et al. 2004. Perfect orientation ordered in-situ one-dimensional self-assembly of Mn-doped PbSe nanocrystals. *J. Am. Chem. Soc.* 126: 14816–21.

Ma, M. G., and Zhu, J. F. 2009. Solvothermal synthesis and characterization of hierarchically nanostructured hydroxyapatite hollow spheres. *Eur. J. Inorg. Chem.* 36: 5522–6.

Ma, M. G., Zhu, Y. J., and Chang, J. 2008. Solvothermal preparation of hydroxyapatite microtubes in water/N,N-dimethylformamide mixed solvents. *Mater. Lett.* 62: 1642–5.

Mandel, S., and Tas, A. C. 2010. Brushite ($CaHPO_4 \cdot 2H_2O$) to octacalcium phosphate ($Ca_8(HPO_4)_2(PO_4)_4 \cdot 5H_2O$) transformation in DMEM solutions at 36.5°C. *Mater. Sci. Eng. C* 30: 245–54.

Martins, M. A., Santos, C., Almeida, M. M., et al. 2008. Hydroxyapatite micro- and nanoparticles: Nucleation and growth mechanisms in the presence of citrate species. *J. Colloid Interf. Sci.* 318: 210–16.

Moradian-Oldak, J. 2001. Amelogenins: Assembly, processing and control of crystal morphology. *Matrix Biol.* 20: 293–305.

Moradian-Oldak, J., Tan, J., and Fincham, A. G. 1998. Interaction of amelogenin with hydroxyapatite crystals: An adherence effect through amelogenin self-association. *Biopolymers* 46: 225–38.

Narayanaswamy, A., Xu, H. F., Pradhan, N., et al. 2006. Formation of nearly monodisperse In_2O_3 nanodots and oriented-attached nanoflowers: Hydrolysis and alcoholysis vs pyrolysis. *J. Am. Chem. Soc.* 128: 10210–9.

Nishikawa, H. 2003. Radical generation on hydroxyapatite by UV irradiation. *Mater. Lett.* 58: 14–16.

Notodihardjo, F., Kakudo, N., Kushida, S., et al. 2012. Bone regeneration with BMP-2 and hydroxyapatite in critical-size calvarial defects in rats. *J. Cranio. Maxill. Surg.* 40: 287–91.

Onuma, K., Kanzaki, N., and Ito, A., et al. 1998. Growth kinetics of the hydroxyapatite (0001) face revealed by phase shift interferometry and atomic force microscopy. *J. Phys. Chem. B* 102: 7833–8.

Pieters, I. Y., Van den Vreken, N. M. F., Declercq, H. A., et al. 2010. Carbonated apatites obtained by the hydrolysis of monetite: Influence of carbonate content on adhesion and proliferation of MC3T3-E1 osteoblastic cells. *Acta Biomater.* 6: 1561–8.

Posner, A. S., and Betts, F. 1975. Synthetic amorphous calcium-phosphate and its relation to bone-mineral structure. *Acc. Chem. Res.* 8: 273–81.

Ramanan, S. R., and Venkatesh, R. 2004. A study of hydroxyapatite fibers prepared via sol-gel route. *Mater. Lett.* 58: 3320–3.

Ribeiro, N., Sousa, S., and Monteiro, F. 2010. Influence of crystallite size of nanophased hydroxyapatite on fibronectin and osteonectin adsorption and on MC3T3-E1 osteoblast adhesion and morphology. *J. Colloid Interf. Sci.* 351: 398–406.

Robinson, A. 1952. An electron microscopic study of the crystalline inorganic components of bone and its relationship to the organic matrix. *J. Bone Jt. Surg.* 34A: 389–434.

Sahoo, N. G., Kakran, M., Shaal, L. A., et al. 2011. Preparation and characterization of quercetin nanocrystals. *J. Pharm. Sci.* 100: 2379–90.

Salarian, M., Solati-Hashjin, M., Shafiei, S. S., et al. 2009. Template-directed hydrothermal synthesis of dandelion-like hydroxyapatite in the presence of cetyltrimethylammonium bromide and polyethylene glycol. *Ceram. Int.* 35: 2563–9.

Schmidt, H., Gray, B., Wingert, P., et al. 2004. Assembly of aqueous-cored calcium phosphate nanoparticles for drug delivery. *Chem. Mater.* 16: 4942–7.

Schmidt, H., and Ostafin, A. 2002. Liposome directed growth of calcium phosphate nanoshells. *Adv. Mater.* 14: 532–5.

Schnell, E., Klinkhammer, K., Balzer, S., et al. 2007. Guidance of glial cell migration and axonal growth on electrospun nanofibers of poly-ε-caprolactone and a collagen/poly-ε-caprolactone blend. *Biomaterials* 28: 3012–25.

Seo, D. S., and Lee, J. K. 2008. Synthesis of hydroxyapatite whiskers through dissolution–reprecipitation process using EDTA. *J. Cryst. Growth* 310: 2162–67.

Shen, Y., Liu, J., Lin, K., et al. 2012. Synthesis of strontium substituted hydroxyapatite whiskers used as bioactive and mechanical reinforcement material. *Mater. Lett.* 70: 76–79.

Spanos, N., Klepetsanis, P. G., and Koutsoukos, P. G. 2001. Model studies on the interaction of amino acids with biominerals: The effect of L-serine at the hydroxyapatite–water interface *J. Colloid Interface Sci.* 236: 260–5.

Spoerke, E. D., Anthony, S. G., and Stupp, S. I. 2009. Enzyme directed templating of artificial bone mineral. *Adv. Mater.* 21: 425–30.

Stenport, V., Kjellin, P., Andersson, M., et al. 2008. Precipitation of calcium phosphate in the presence of albumin on titanium implants with four different possibly bioactive surface preparations. An *in vitro* study. *J. Mater. Sci: Mater. Med.* 19: 3497–505.

Su, X., Sun, K., and Cui, F. 2003. Organization of apatite crystals in human woven bone. *Bone* 32: 150–62.

Tabrizi, B., Honarmandi, P., Kahrizsangi, R., et al. 2009. Synthesis of nanosize single-crystal hydroxyapatite via mechanochemical method. *Mater. Lett.* 63: 543–6.

Tao, J., Pan, H., Zeng, Y., et al. 2007. Roles of amorphous calcium phosphate and biological additives in the assembly of hydroxyapatite nanoparticles. *J. Phys. Chem. B* 111: 13410–8.

Taylor, H. F. W. 1986. Proposed structure for calcium siliate hydrate gel. *J. Am. Ceram. Soc.* 69: 464–7.

Tjandra, W., Ravi, P., Yao, J., et al. 2006. Synthesis of hollow spherical calcium phosphate nanoparticles using polymeric nanotemplates. *Nanotechnology* 17: 5988–94.

Tornblom, M., and Henriksson, U. 1997. Effect of solubilization of aliphatic hydrocarbons on size and shape of rodlike C_{16}TABr micelles studied by 2H NMR relaxation. *J. Phys. Chem. B* 101: 6028–35.

Toyasawa, S., Ohuigin, C., Figueroa, F., et al. 1998. Identification and characterization of amelogenin genes in monotremes, reptiles, and amphibians. *Proc. Natl. Acad. Sci. USA* 95: 13056–61.

Viswanath, B., and Ravishankar, N. 2008. Controlled synthesis of plate-shaped hydroxyapatite and implications for the morphology of the apatite phase in bone. *Biomaterials* 29: 4855–63.

Wakae, H., Takeuchi, A., Udoh, K., et al. 2008. Fabrication of macroporous carbonate apatite foam by hydrothermal conversion of a-tricalcium phosphate in carbonate solutions. *J. Biomed. Mater. Res. A.* 87: 957–63.

Wang, K. W., Zhu, Y. J., Chen, F., et al. 2011. Microwave-assisted synthesis of hydroxyapatite hollow microspheres in aqueous solution. *Mater. Lett.* 65: 2361–3.

Wang, K. W., Zhu, Y. J., Chen, X. Y., et al. 2010. Flower-like hierarchically nanostructured hydroxyapatite hollow spheres: Facile preparation and application in anticancer drug cellular delivery. *Chem. Asian J.* 5: 2477–82.

Wang, L., Guan, X., Du, C., et al. 2007. Amelogenin promotes the formation of elongated apatite microstructures in a controlled crystallization system. *J. Phys. Chem. C* 111: 6398–404.

Wang, Y. J., Chen, J. D., Wei, K., et al. 2006. Surfactant-assisted synthesis of hydroxyapatite particles. *Mater. Lett.* 60: 3227–31.

Wang, Y. J., Lai, C., Wei, K.; Chen, X. F. et al. 2006. Investigations on the formation mechanism of hydroxyapatite synthesized by the solvothermal method. *Nanotechnology* 17: 4405–12.

Wang, Y. S., Hassan, M. S., Gunawan, P., et al. 2009. Polyelectrolyte mediated formation of hydroxyapatite microspheres of controlled size and hierarchical structure. *J. Colloid Interf. Sci.,* 339: 69–77.

Wei, K., Lai, C., and Wang, Y. J. 2006. Solvothermal synthesis of calcium phosphate nanowires under different pH condition. *J. Macromol. Sci. A.* 43: 1531–40.

Weiner, S., Arad, T., and Traub, W. 1991. Crystal organization in rat bone lamellae. *FEBS Lett.* 285: 49–54.

Weiner, S., and Wagner, H. 1998. The material bone: Structure mechanical function relations. *Annu. Rev. Mater. Sci.* 28: 271–98.

Wen, H. B., Moradian-Oldak, J., and Fincham, A.G. 2000. Dose dependent modulation of octacalcium phosphate crystal habit by amelogenins. *J. Dent. Res.* 79: 1902–6.

Wong, M., Eulenberger, J., Schenk, R., et al. 1995. Effect of surface-topology on the osseointegration of implant materials in trabecular bone. *J. Biomed. Mater. Res.* 29: 1567–75.

Wu, L., Dou, Y., Lin, K. 2011. Hierarchically structured nanocrystalline hydroxyapatite assembled hollow fibers as a promising protein delivery system. *Chem. Commun.* 47: 11674–6.

Wu, Y., Hench, L. L., Du, J., et al. 2004. Preparation of hydroxyapatite fibers by electrospinning technique. *J. Am. Ceram. Soc.* 87: 1988–91.

Xiao, X., Liu, R., Qiu, C., et al. 2009. Biomimetic synthesis of micrometer spherical hydroxyapatite with β-cyclodextrin as template. *Mater. Sci. Eng. C* 29: 785–90.

Xie, J., Blough, E. R., and Wang, C. H. 2012. Submicron bioactive glass tubes for bone tissue engineering. *Acta Biomater.* 8: 811–9.

Yamaguchi, T., Sakai, S., and Kawakami, K. 2008. Application of silicate electrospun nanofibers for cell culture. *J. Sol-Gel Sci. Technol.* 48: 350–5.

Yang, S., He, H., Wang, L., et al. 2011. Oriented crystallization of hydroxyapatite by the biomimetic amelogenin nanospheres from self-assemblies of amphiphilic dendrons. *Chem. Commun.* 47: 10100–2.

Ye, F., Guo H., and Zhang, H. 2008. Biomimetic synthesis of oriented hydroxyapatite mediated by nonionic surfactants. *Nanotechnology* 19: 245605.

Yokoi, T., Kawashita, M., Kikuta, K., et al. 2010. Biomimetic mineralization of calcium phosphate crystals in polyacrylamide hydrogel: Effect of concentrations of calcium and phosphate ions on crystalline phases and morphology. *Mater. Sci. Eng. C* 30: 154–9.

Yuan, Y., Liu, C., Zhang, Y., et al. 2008. Sol-gel auto-combustion synthesis of hydroxyapatite nanotubes array in porous alumina template. *Mater. Chem. Phys.* 112: 275–80.

Zhang, C. M., Cheng, Z. Y., Yang, P. P., et al. 2009. Architectures of strontium hydroxyapatite microspheres: Solvothermal synthesis and luminescence properties. *Langmuir* 25: 13591–8.

Zhang, C. M., Yang, J., Quan, Z. W., et al. 2009. Hydroxyapatite nano- and microcrystals with multiform morphologies: Controllable synthesis and luminescence properties. *Cryst. Growth Des.* 9: 2725–33.

Zhang, H., and Darvell, B. W. 2010a. Constitution and morphology of hydroxyapatite whiskers prepared using amine additives. *J. Eur. Ceram. Soc.* 30: 2041–8.

Zhang, H., and Darvell, B. W. 2010b. Synthesis and characterization of hydroxyapatite whiskers by hydrothermal homogeneous precipitation using acetamide. *Acta Biomater.* 6: 3216–22.

Zhang, P., and Gao, L. 2003. Cadmium sulfide nanocrystals via two-step hydrothermal process in microemulsions: Synthesis and characterization. *J. Colloid Interf. Sci.* 266: 457–60.

Zhang, Y., and Lu, J. 2007. A simple method to tailor spherical nanocrystal hydroxyapatite at low temperature. *J. Nanopart. Res.* 9: 589–94.

Zou, B., Liu, Y., Luo, X., et al. 2012. Electrospun fibrous scaffolds with continuous gradations in mineral contents and biological cues for manipulating cellular behaviors. *Acta Biomater.* 8: 1576–85.

7

Bioactive Inorganic and Organic Composite Materials for Bone Regeneration and Gene Delivery

Yufeng Zhang and Chengtie Wu

CONTENTS

7.1 Introduction

Bone defects and malformation, caused by trauma, infection, tumor resection, congenital deformity, and physical and pathological degeneration, represent a major concern for orthopedic surgeons. According to the Office of the U.S. Surgeon General, bone diseases affect 10 to 12 million people in the United States and there are approximately 1.5 million fractures reported in the United States every year (Shane 2010). Within the European Union, it is estimated that a bone fracture occurs every 30 seconds as a consequence of osteoporosis (Compston 1999) and in China over 3 million individuals per year are estimated to suffer from bone defects or injury; figures that only superficially describe the enormous impact of such injuries on the quality of life of individuals and health care costs carried by society (Kesheng et al. 2006). Considering this data, we can see that bone repair and regeneration can solve a serious problem for human beings.

Bone tissues have a native potential for self-healing; however, many patients, especially those with critical size defects or poor bone quality, still require surgical treatments. However, due to lack of donor tissue and possible donor site morbidity, surgical treatment is not a perfect choice. Bone tissue engineering represents a promising alternative to generate new bone for any desired size and shape, and is the current therapeutic concept for bone defect healing. One of the commonly accepted definitions of tissue engineering was proposed by Langer and Vacanti in 1993: "an interdisciplinary field that applies the principles of engineering and life sciences toward the development of biological substitutes that restore, maintain, or improve tissue function or a whole organ." Tissue engineering is a multidisciplinary field in which physicians, engineers, and scientists seek to provide novel solutions for orthopedic surgery in an effort to overcome the previously mentioned disadvantages. The extensively recognized key elements for bone tissue engineering and morphogenesis are osteogenic cells, osteoconductive scaffolds or extracellular matrices (ECM), and osteoinductive growth factors under mechanical microenvironment (Figure 7.1).

Great progress has been made using bioactive materials for bone tissue repair and regeneration has as a result of the scientific efforts aimed at improving the tissue–material response after implantation. The extent of bone regeneration could be enhanced by using appropriate bioactive materials. According to the present classifications, we can generally separate it into inorganic and organic materials. Inorganic biomaterials such as hydroxyapatite, tricalcium phosphate, bioactive glasses, and silicate ceramics produce tenacious bonds with hard tissue by reacting with physiologic fluids. However, stiff and brittle characteristics make it difficult to form into complex shapes. On the other side, composite polymers are easily fabricated into complex structures, but they are too weak to meet the demands of surgery and the *in vivo* physiologic environment (Boccaccini and Blaker 2005).

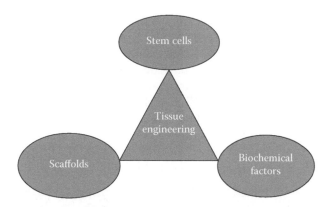

FIGURE 7.1
The three main pillars that support bone tissue engineering: cells, scaffolds, and biochemical factors.

Therefore, people are thinking about the compounds of the inorganic and organic materials to solve its problems and to strengthen its benign properties. First, the addition of bioactive phases to polymers offers a pH buffering effect at the polymer surface, thus acceleration of acidic degradation of the polymer can be controlled. Second, by control of the structure and volume of fraction, the composite materials can avoid the erosion of degradation. Third, composite materials are able to absorb water due to the internal interfaces formed between the polymer and more hydrophilic bioactive phases.

Gene therapy has become one of the most promising methods to treat disease by delivering nucleotide sequence, principally DNA and various types of RNA including antisense RNA, siRNA, isRNA, and miRNA, into a specific cell population, thereby expressing successfully to manipulate cellular activity, which in turn synthesizes functional proteins to stimulate immune responses or tissue regeneration, or blocks expression at the level of transcription or translation for treatment of several diseases (Piskin et al. 2004; Wagner and Bhaduri 2012). The classic approach to gene therapy consists of three elements: (1) an expression gene coding a specific therapeutic protein, (2) target cells such as stem cells or precursor cells, and (3) a gene delivery system that delivers the gene into the target cells (Schaffer and Lauffenburger 1998; Piskin et al. 2004). The gene delivery system is one of the most important elements of the therapeutic performance required of these features including high transfection efficacy, desirable biodistribution, appropriate degradation rate, and bioactivity to facilitate the intracellular trafficking of the gene expression system (Sundaram et al. 2009).

Traditionally, there are two different strategies to effectively deliver the target gene to the proper cells or tissues with a vector or without a vector (Luo and Saltzman 2000). The gene delivery system without a vector (mainly by physical methods) is difficult to control and optimize *in vivo* restricting its study and application (Kamimura et al. 2011). On the other hand, most gene therapies use vectors (viral and nonviral) to enhance the transfection and expression of the target gene (Scaduto and Lieberman 1999). Different viruses, with adenoviruses, retroviruses, lentiviruses, and adeno-associated viruses (AAVs) among the most common, have been introduced as gene delivery vectors and applied extensively (Kootstra and Verma 2003; Roth and Sundaram 2004; Zhang and Godbey 2006; Zhang et al. 2009) (Figure 7.2). The nonviral vectors divided into organic and inorganic mimic functions of viral transfection but eschew side effects of viral vectors (Wagner and Bhaduri 2012).

The organic and inorganic each have its own merits. The characteristics of organics, such as lipids, polycations, and intelligent polymers, are exploited

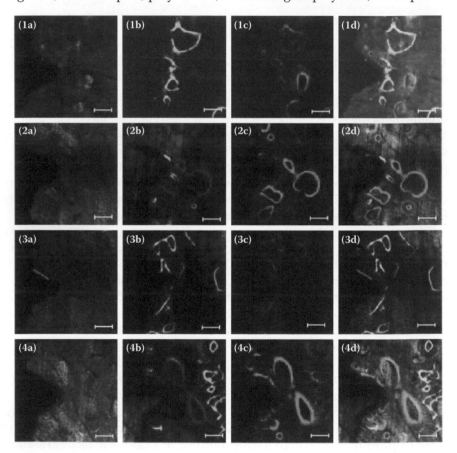

FIGURE 7.2
(**See color insert.**) New bone formation was determined histomorphometrically by bone marker quantification, representing the different healing periods in the four groups, which are enumerated as respective rows in the following matrix of images. Column (a) shows the basic confocal LASER microscope image of samples. Column (b) shows fluorescence of calcein introduced 4 weeks after implantation. Column (c) shows fluorescence of alizarin introduced 8 weeks after implantation. Column (d) shows folded images of the two fluorochromes with the basic confocal image.

to concentrate the DNA as a result of dissimilar charges attract intensify permeability by surface modification and biodegrade into low-toxic product (Hwang and Davis 2001; Ratner and Bryant 2004). Meanwhile, its fast degradation and weak mechanical strength impose restrictions on its application especially for load bearing (Di Martino et al. 2005). Following the research progresses, inorganic materials have shown comparable and, in some instances, better gene delivery efficiencies than organic systems (Sokolova and Epple 2008). Calcium phosphates (CaPs), gold nanoparticles (GNPs), silica, magnesium phosphates, and iron oxides have been introduced for study and exhibited excellent characteristics lying in antibacterial activity, biocompatibility, and bioactivity (Kim et al. 1998; Chowdhury and Akaike 2005; Kawano et al. 2006). However, weak interactions between the inorganic vector and the ribonucleic acid molecules are usually established in the case of gene delivery system (Vallet-Regi et al. 2011).

In recent years, research into bioactive organic–inorganic hybrid materials has captured the scientific and public interest with the promise to gene delivery. Bioactive organic–inorganic hybrid materials can be generally defined as materials with organic and inorganic components that are not only closely mixed in phases but also coexist intimately through size domain effects and nature of the interfaces (Gómez-Romero and Sanchez 2004; Vallet-Regi et al. 2011). Bioactive organic–inorganic hybrid materials offer unique biological, electrical, and mechanical properties for the design of an excellent gene delivery system (Yang et al. 2009).

For the sake of organization, this chapter has been divided into two parts of bone regeneration and gene delivery. Each part presents three classifications that are the combination of CaP inorganic materials including phosphate (hydroxyapatite [HAp], tricalcium phosphate [TCP], or biphasic calcium phosphate [BCP]) composite materials, the bioactive glass composite materials, and the silicate bioceramic composite materials. We will discuss the microstructure and mechanical properties of these materials, then evaluate the feasibility in the field of biomedical materials. Furthermore, three sections were organized to introduce the gene delivery of composites materials, including CaP-based hybrid materials, hybrid materials integrated with gold nanoparticles (GNPs), silicate-based hybrid materials, quantum-dots-based hybrid materials, and iron-based hybrid materials for gene delivery.

7.2 Bioactive Inorganic–Organic Composite Materials for Bone Regeneration

Tissue engineering composites are potential bone substitutes. The brittleness of inorganic materials is their main disadvantage as scaffold materials. Polymers are easily fabricated to form complex shapes and structures, but they lack a bioactive function (e.g., strong bonding to living tissue) and are too weak to meet the mechanical requirement in surgery and in the physiologic environment. Therefore, it is a viable way to combine biodegradable polymers and bioactive inorganic materials for tissue engineering applications.

7.2.1 Composites of Calcium Phosphate Bioceramics and Polymers

Calcium phosphate biomaterials are conventional materials widely used in bone repair and regeneration. The next generation of calcium phosphate biomaterials should combine bioactive and bioresorbable properties to activate *in vivo* mechanisms of tissue regeneration to stimulate the body to heal itself and lead to replacement of the scaffold by the regenerating tissue. Here we mainly discussed three kinds of calcium phosphate materials—HAp, TCP, and BCP—to review CaP composite materials in bone regeneration. It was recently found that these biodegradable polymers have already been combined with inorganic biomaterials for bone tissue engineering, such as polylactide (PLA), polyglycolic acid (PGA), polylactid-co-glycolic acid (PLGA), poly-caprolactone (PCL), poly (lactide-ε-caprolactone) (PLCL), platelet-rich fibrin (PRF), polyethylene oxide (PEO), poly(3-hydroxy-butytrate) (PHB), silk, hyaluronic acid, alginate, chitosan, fibrin, and collagen.

7.2.1.1 Hydroxyapatite (HAp)

7.2.1.1.1 Particle

Kim (2007) prepared a HAp–PCL nanocomposite that can distribute HAp nanoparticles uniformly within the PCL matrix in the mediation of oleic acid. Wang, Chow, et al. (2011) applied a calcium-phosphate/phosphonate hybrid shell to introduce a greater amount of reactive hydroxyl groups onto the HAp particles. PLA-grafted HAp shows significantly different pH-dependent ζ-potential and particle size profiles from those of uncoated HAp. The diametric tensile strength of PLA–HAp composite prepared from PLA-grafted HAp was found to be over twice that of the composite prepared from the nonmodified HAp. Moreover, the tensile strength of the improved composite was 23% higher than that of PLA alone (Wang, Chow, et al. 2011). Gaharwar et al. (2011) presented poly(ethylene glycol) PEG and hydroxyapatite nanoparticles (nHAp nanoparticles). They have higher extensibilities, fracture stresses, compressive strengths, and toughness when compared with conventional PEO hydrogels. The combination of PEG and HAp nanoparticles significantly improved the physical and chemical hydrogel properties as well as some biological characteristics, such as osteoblast cell adhesion (Gaharwar et al. 2011).

Comparing collagen (COL)-fibroin/hydroxyapatite (COL-SF/HAp) composite, COL-HAp, and SF-HAp, respectively, Wang, Zhou, et al. (2011) found that the mineral phase in COL-SF/HAp displayed smaller size and more narrow distribution. Of all the above biomimetic composites, the HAp was well assembled with molecular template(s), and the organic content of the composite was about 12% to 20%, which was quite similar to the natural bone in composition. The results revealed that the spatial structure of coassembly template proteins played a pivotal role in controlling and regulating HAp crystal nucleation and growth (Wang et al. 2011).

Preosteoblast MC3T3-E1 cells were cultured in hyaluronic acid-modified chitosan/collagen/nano-hydroxyapatite (HAp-CS/Col/nHAp) composite scaffolds and treated with phytoestrogen α-zearalanol (α-ZAL) to improve bone tissue formation for bone tissue engineering. Osteogenic phenotype increased or maintained enhanced collagen I (COLI) levels, decreased osteopontin expression, and had little effect on osteocalcin expression during 12 days of *in vitro* culture. In response to α-ZAL, the cell-scaffold constructs inhibited cellular proliferation, enhanced the alkaline phosphatase (ALP) activity, and increased the ratio of osteoprotegerin to receptor activator of nuclear factor kappa B (NF-κB) ligand (RANKL). Application of perfusion and dynamic strain to cell-scaffold constructs treated with α-ZAL represents a promising approach in the studies of osteogenesis stimulation of bone tissue engineering (Liu et al. 2012).

The achievement of nanodistribution for inorganic reinforced filler is a significant challenge to three-dimensional porous composite scaffolds. A homogeneous nano hydroxyapatite/polyelectrolyte complex (HAp/PEC) hybrid scaffold was developed and investigated. Based on the enhancing properties of the formation of PEC between chitosan and hyaluronic acid, the biocompatibility and bioactivity were evaluated by human bone mesenchymal stem cells (hBMSCs) proliferation (MTT assay), differentiation (alkaline phosphatase [ALP] activity), and histological analysis. The *in vitro* tests show that the scaffold is an excellent carrier for cell penetration, growth, and proliferation for bone repair application (Chen et al. 2012).

7.2.1.1.2 Microspheres

Peng et al. (2010) reported a large scaffold 3 to 4 cm in length and 1 to 1.5 cm in diameter designed for engineering large bone tissue *in vivo*. The scaffold was made by filling HAp spherules, prepared by chitin sol emulsification in oil and gelation in situ, into a porous HAp tube coated with a thin layer of poly(L-lactic acid) (PLA) (Peng et al. 2010). Jevtic et al. (2009) obtained PLGA/HAp biocomposite nanospheres via an ultrasound way. Optimal parameters for the formation of PLGA/HAp included a lower content of the ceramic phase (PLGA/HAp = 90:10), higher power of ultrasonic field (P = 142.4W), lower temperature of the medium during ultrasonic treatment (T = 8°C), dilute solution of PVP as surfactant and dispersion of HAp in polymer solution in order to achieve the desired homogeneity before the formulation of the composite. The morphology of PLGA/HAp particles was highly uniform and spher-like, with small dimensions (150–320 nm), highly uniform particle size distribution, and characteristics of planar spatial self-organization (Jevtic et al. 2009).

7.2.1.1.3 Scaffolds

Kothapalli et al. (2005) fabricated a kind of scaffold comprising poly(D,L-lactic acid) (PLA) and nano-hydroxyapatite (HAp) prepared by using the solvent-casting/salt-leaching technique. The particles had an average size of approximately 25 nm in width and 150 nm in length with aspect ratios ranging from 6 to 8. As the HAp content increased in the scaffold from 0 to 50 wt%, the compression modulus of the scaffolds increased from 4.72±1.2 to 9.87±1.8 MPa, while the yield strength increased from 0.29±0.03 to 0.44±0.01 MPa (Kothapalli et al. 2005). Son et al. (2011) reported a combination of a ceramic/polymer biphasic scaffold. That is, the outside cortical-like shells, composed of porous HAp with a hollow interior using a polymeric template-coating technique and the inner trabecular-like core which consisted of porous poly(D,L-lactic acid) (PLA) was loaded with dexamethasone (DEX) and was directly produced using a particle leaching/gas forming technique to create the inner diameter of the HAp scaffold (Son et al. 2011). Causa et al. (2006) discovered HAp particles-PCL scaffold has the potential to be an efficient substrate for bone substitution after the assessment of cell viability, proliferation, morphology, and ALP. HAp-loaded PCL was found to improve osteoconduction compared to the PCL alone in the condition of Saos-2 cells and osteoblasts from human trabecular bone (hOB) retrieved during total hip replacement surgery seeded onto 3D PCL samples for 1 to 4 weeks (Causa et al. 2006). Guarino et al. (2008) described a fiber-reinforced composites scaffold composed of poly-L-lactide acid (PLLA) fibers embedded in a porous poly(epsilon-caprolactone) matrix. The porosity degree can reach 79.7% and the bimodal pore size distribution is in the range of 10 to 200 μm (Guarino et al. 2008).

A novel combination of polyurethane (PU) foam method and a hydrogen peroxide (H_2O_2) foaming method is used to fabricate the macroporous HAp scaffolds. The internal surfaces of the macropores are further coated with a poly(D,L-lactic-co-glycolic acid) PLGA-bioactive glass composite coating. It is found that the HAp scaffolds fabricated by the combined method show high porosities of 61% to 65% and proper macropore sizes of 200 to 600 μm. The PLGA infiltration improved the compressive strengths of the scaffolds from 1.5–1.8 to 4.0–5.8 MPa. Furthermore, the bioactive glass-PLGA coatings rendered a good bioactivity to the composites, evidenced by the formation of an apatite layer on the sample surfaces immersed in the simulated body fluid (SBF) for 5 days (Huang and Miao 2007). Xu et al. (2012) compared two kinds of vascular stents. In this study, vascular stents were fabricated from poly (lactide-ε-caprolactone)/collagen/nano-hydroxyapatite (PLCL/Col/nHAp) by electrospinning. In addition, nanocomposite scaffolds of poly (lactic-co-glycolic acid)/polycaprolactone/nano-hydroxyapatite (PLGA/PCL/nHAp) loaded with the vascular stents were prepared by thermoforming-particle leaching and their basic performance and osteogenesis were tested *in vitro* and *in vivo*. The results showed that the PLCL/Col/nHAp stents and PLGA/

FIGURE 7.3
SEM micrographs for the (a) pure CaP powders and (b) hybrid CaP/silk powders.

PCL/nHAp nanocomposite scaffolds had good surface structures, mechanical properties, biocompatibility, and osteoconductivity (Xu et al. 2012).

Zhang et al. (2009) investigated a porous CaP/silk composite scaffolds via a freeze-drying method to facilitate osteogenic properties of human bone mesenchymal stromal cells (BMSCs) and *in vivo* bone formation abilities. The results showed that incorporating the hybrid CaP/silk powders into silk scaffolds improved both pore structure architecture and distribution of CaP powders in the composite scaffolds. And *in vitro* osteogenic differentiation of BMSCs was enhanced and cancellous bone formation was increased (Figure 7.3) (Zhang et al. 2010).

7.2.1.2 Tricalcium Phosphate (TCP)

β-tricalcium phosphate (β-TCP) has been widely used to regenerate various hard tissues. Although bioceramics and collagen have various biological advantages with respect to cellular activity, their usage has been limited due to the inherent brittleness and low mechanical properties of β-TCP, along with the low shapeability of the three-dimensional collagen. Therefore, there are many studies based on β-TCP to produce different composites to improve its characteristics.

7.2.1.2.1 Scaffolds

Haimi et al. (2009) compared the effects of novel three-dimensional composite scaffolds consisting of a bioactive phase (bioactive glass o rβ-TCP 10 and 20 wt%) incorporated within a polylactic acid (PLA) matrix on viability, distribution, proliferation, and osteogenic differentiation of human adipose stem cells (ASCs). DNA content and ALP activity of ASCs cultured on PLA/β-TCP composite scaffolds were higher than other types of scaffolds. Interestingly, the cell number was significantly lower, but the relative ALP/

DNA ratio of ASCs was significantly higher in PLA/bioactive glass scaffolds than in the other three scaffold types. These results indicate that the PLA/β-TCP composite scaffolds significantly enhance ASC proliferation and total ALP activity compared to other scaffold types (Haimi et al. 2009).

Yanoso-Scholl et al. (2010) investigated the microstructural and mechanical properties of dense PLA and PLA/beta-TCP (85:15) scaffolds fabricated using a rapid volume expansion phase separation technique, which embeds uncoated β-TCP particles within the porous polymer. PLA scaffolds had a volumetric porosity in the range of 30% to 40%. With the embedding of β-TCP mineral particles, the porosity of the scaffolds decreased in half, whereas the ultimate compressive and torsion strength were significantly increased. When loaded with BMP2 and VEGF and implanted in the quadriceps muscle, PLA/beta-TCP scaffolds did not induce ectopic mineralization but induced a significant 1.8-fold increase in neovessel formation (Yanoso-Scholl et al. 2010). The biocomposite (PLA/β-TCP) was compared with a currently used β-TCP bone substitute (ChronOS, Dr. Robert Mathys Foundation) representing a positive control, and empty defects representing a negative control. Ten defects were created in sheep cancellous bone, three in the distal femur, and two in the proximal tibia of each hind limb, with diameters of 5 mm and depths of 15 mm. Bone ingrowth was observed in the biocomposite scaffold, including its central part. Despite initial fibrous tissue formation observed at 2 and 4 months, but not at 12 months, this initial fibrous tissue does not preclude long-term application of the biocomposite, as demonstrated by its osteointegration after 12 months, as well as the absence of chronic or long-term inflammation at this time point (Van der Pol et al. 2010). Haaparanta et al. (2010) applied freeze-drying fabricating porous PLA/β-TCP composite scaffolds to characterize these graded porous composite scaffolds in two different PLA concentrations (2 and 3 wt%). Also, three different β-TCP ratios (5, 10, and 20 wt%) were used to study the effect of β-TCP on the properties of the polymer. The characterization was carried out by determining the pH, weight change, component ratios, thermal stability, inherent viscosity, and microstructure of the scaffolds in 26 weeks of hydrolysis. We observed that the fabrication method improved the thermal stability of the samples. The dense surface skin of the scaffold may inhibit the ingrowth of osteoblasts and bone tissue, while simultaneously stimulating the ingrowth of chondrocytes (Haaparanta et al. 2010). McCullen et al. (2009) fabricated electrospun composite scaffolds consisting of β-TCP crystals and poly(L-lactic acid) (PLA) at varying loading levels of β-TCP (0, 5, 10, 20 wt%). With the addition of β-TCP, the fiber diameter increased with each treatment ranging from 503.39±20.31 nm for 0 wt% β-TCP to 1267.36±59.03 nm for 20 wt% β-TCP. The overall tensile strength of the neat scaffold (0 wt% β-TCP) was 847±89.43 kPa; the addition of β-TCP significantly decreased this value to an average of 350.83±38.57 kPa. DNA content increased in a temporal manner for each scaffold over 18 days in culture, although day 12 and the 10 wt% TCP scaffold induced the greatest hASC proliferation. Endogenous

ALP activity was enhanced on the composite PLA/TCP scaffolds compared to the PLA control particularly by day 18. It was noted that at the highest β-TCP loading levels of 10 and 20 wt%, there was a dramatic increase in the amount of cell-mediated mineralization compared to the 5 wt% TCP and the neat PLA scaffold (McCullen et al. 2009). Yeo et al. (2011) demonstrated greater degradation of PCL-TCP scaffolds *in vivo* than *in vitro*. At 24 weeks, the increase of average porosity of the scaffolds *in vivo* was 29.2% compared to 2.65% *in vitro*. Gel permeation chromatography (GPC) analysis revealed a decrease of 29% and 20%, respectively, in the Mn and Mw values after 24 weeks *in vitro*. However, a significant decrease in Mn and Mw values (79.6% and 88.7%, respectively) were recorded *in vivo*. The mechanical properties closely match those of cancellous bone even at 24 weeks. The results showed that the scaffold can be used for dentoalveolar reconstruction and PCL-TCP scaffolds have shown to possess the potential to degrade within the desired time period of 5 to 6 months and favorable mechanical properties (Lee et al. 2008). Yeo et al. (2011) fabricated a new hierarchical scaffold that consisted of a melt-plotted polycaprolactone (PCL)/β-TCP composite and embedded collagen nanofibers. The fabrication process was combined with general melt-plotting methods and electrospinning. Scanning electron microscope (SEM) micrographs of the fabricated scaffolds indicated that the β-TCP particles were uniformly embedded in PCL struts and that electro-spun collagen nanofibers (diameter = 160 nm) were well layered between the composite struts. By accommodating the β-TCP and collagen nanofibers, the hierarchical composite scaffolds showed dramatic water-absorption ability (100% increase), increased hydrophilic properties (20%), and good mechanical properties. MTT assay and SEM images of cell-seeded scaffolds showed that the initial attachment of osteoblast-like cells (MG63) in the hierarchical scaffold was 2.2 times higher than that on the PCL/β-TCP composite scaffold. Additionally, the proliferation rate of the cells was about 2 times higher than that of the composite scaffold after 7 days of cell culture (Yeo et al. 2011). Cao and Kuboyama (2010) applied the solvent casting and particulate leaching method to study composite scaffolds of PGA/beta-TCP (in 1:1 and 1:3 weight ratios) in the repair of critical bone defects (3 mm diameter, 2 mm depth) in rat femoral medial-epicondyles, compared with HAp and blank controls. The results showed that the new bone mineral densities (mg/cm^3) with HAp, PGA/β-TCP (1:1), and PGA/β-TCP (1:3) at 90 days after surgery were 390.4±18.1, 563.8±26.9, and 606.3±26.9, respectively. The biodegradation percents (%) of HAp, PGA/β-TCP (1:1), and PGA/β-TCP (1:3) at 90 days after surgery were 35.1±−5.5, 99.0±−1.0, and 96.2±−3.3, respectively. The PGA/β-TCP scaffolds were almost replaced by new growing bone within 90 days after surgery. Thus, the PGA/β-TCP composite scaffold, especially with a weight ratio 1:3, exhibited a strong ability for osteogenesis, mineralization, and biodegradation for bone replacement (Cao and Kuboyama 2010). Kim et al. (2012) prepared poly(d,l-lactide:glycolide) (DL-PLGA) and β-TCP nanocomposites via fused deposition modeling (FDM), a type of extrusion free-form fabrication. Microfilaments deposited at angles of 0° and 90° were designated as the "simple" scaffold architecture, whereas those deposited at angles alternating between 0°, 90°, 45°, and −45° were designated as the "complex" scaffold architecture. The scaffolds were implanted into rabbit femoral unicortical bone defects according to four treatment groups based on pore structure and HAp coating. After implanted for 6 and 12 weeks, scaffolds and host bone were recovered and processed for histology. The results suggest that all configurations of the scaffolds integrated with the host bone were biocompatible and thus might offer an exciting new scaffold platform for delivery of biologicals for bone regeneration (Kim et al. 2012).

Luvizuto et al. (2011) studied the PLLA/PGA/β-TCP biocomposite's osteo-conductivity *in vivo*. The PLLA/PGA/β-TCP biocomposite interference screw completely degraded, and no remnant was present 3 years after implantation for a bone-patellar tendon-bone graft ACL reconstruction. Osteoconductivity was confirmed in 21 of 26 screw sites (81%) and completely filled the site in 5 of 26 (19%) (Luvizuto et al. 2011). Reichert et al. (2012) used ovine tibial defects as the experiment mode to search the regenerative potential of medical grade polycaprolactone-tricalcium phosphate (mPCL-TCP) and silk-hydroxyapatite (silk-HAp) scaffolds. Defects were left untreated, then reconstructed with autologous bone grafts (ABG) and mPCL-TCP or silk-HAp scaffolds. Animals were observed for 12 weeks. X-ray analysis, torsion testing, and quantitative computed tomography (M-CT) analyses were performed. The results of this study suggest that mPCL-TCP scaffolds combined can serve as an alternative to autologous bone grafting in long bone defect regeneration. The combination of mPCL-TCP with osteogenic cells or growth factors represents an attractive means to further enhance bone formation (Reichert et al. 2012).

Silk fibroin has been widely used for biomaterial studies by virtue of its combination of mechanical properties, controllable biodegradability, and cytocompatibility. As a material for uses within bone tissue engineering material, 3D porous silk scaffolds are receiving more attention because of their excellent physiochemical properties (Zhang et al. 2010). Bhumiratana et al. (2011) introduced a HAp-silk fibroin scaffold composite to regenerate the bone defects. The result shows that the cultivation of hMSCs in the silk/HAp composite scaffolds under perfusion conditions led to the formation of bone-like structures and an increase in the equilibrium of Young's modulus (up to fourfold or eightfold over 5 or 10 weeks of cultivation, respectively) in a manner that correlated with the initial HAp contents (Bhumiratana et al. 2011).

7.2.1.3 Biphasic Calcium Phosphate (BCP)

7.2.1.3.1 Particles

Roohani-Esfahani et al. (2011) coated the struts of a BCP scaffold with a nanocomposite layer consisting of bioactive glass nanoparticles (nBG) and polycaprolactone (PCL) (BCP/PCL-nBG) to enhance its mechanical and biological behavior. The effect of various nBG concentrations (1–90 wt%) on the mechanical properties and *in vitro* behavior of the scaffolds was comprehensively examined and compared with that for (BCP/PCL-nHAp) and (BCP/PCL). Introduction of 1 to 90 wt% of nBG led to scaffolds with compressive strengths in the range of 0.2 to 1.45 MPa and moduli in the range of 19.3 to 49.4 MPa. This trend was also observed for BCP/PCL-nHAp scaffolds; however, nBG induced even better bioactivity and a faster degradation rate. The maximum compressive strength and modulus increased significantly when 30 wt% nBG was added, compared with BCP scaffolds. Moreover, BCP/PCL-nBG scaffolds induced the differentiation of primary human bone-derived cells (HOBs), with significant upregulation of osteogenic gene expression for Runx2, osteopontin, and bone sialoprotein, compared with the other groups (Roohani-Esfahani et al. 2011).

7.2.1.3.2 Scaffolds

Roohani-Esfahani et al. (2010) reported a composite BCP scaffold by coating a nanocomposite layer, consisting of HAp nanoparticles and PCL, over the surface of BCP. M-CT studies showed that the prepared scaffolds were highly porous (approximately 91%) with large pore size (400–700 μm) and an interconnected porous network of approximately 100%. The HAp nanoparticle (needle shape)-composite coated scaffolds displayed the highest compressive strength (2.1±0.17 MPa), compared to pure HAp/beta-TCP (0.1±0.05 MPa) and to the micron HAp-composite coated scaffolds (0.29±0.07 MPa). They predicted that these properties were of importance for bone regeneration (Roohani-Esfahani et al. 2010). Roohani-Esfahani et al. (2012) present a composite silk/PCL/BCP scaffold that is highly porous (85% for porosity, 500 μm for pore size, and 100% for interconnectivity). Their results showed that using silk only (irrespective of concentration) for the modification of ceramic scaffolds could drastically reduce the compressive strength of the modified scaffolds in aqueous media, and the modification made a limited contribution to improving scaffold toughness. Using PCL/nanostructured silk the compressive strength and modulus of the modified scaffolds reached 0.42 MPa (compared with 0.07 MPa for BCP) and 25 MPa (compared with 5 MPa for BCP), respectively. The failure strain of the modified scaffold increased more than 6% compared with a BCP scaffold (failure strain of less than 1%), indicating a transformation from brittle to elastic behavior. The cytocompatibility of ECM-like composite scaffolds was investigated by studying the attachment, morphology, proliferation, and bone-related gene expression of primary human bone-derived cells. Cells cultured on the developed scaffolds for 7 days had significant upregulation of cell proliferation (1.6-fold higher, P < 0.001) and osteogenic gene expression levels (collagen type I, osteocalcin, and bone sialoprotein) compared with the other groups tested (Roohani-Esfahani et al. 2012). Lu et al. (2012) produced nanocomposite scaffold (BCP/PCL-nHAp) and mimicked the biological microenvironment of bone by coculturing with primary human osteoblasts (HOBs), and then investigated their effects on osteogenic differentiation of adipose tissue-derived stem cells (ASCs). In comparison with the ASCs cultured alone on BCP/PCL, early osteogenic differentiation of ASCs was induced by either seeding ASCs on BCP/PCL-nHAp scaffolds or by coculturing with HOBs; the combination of BCP/PCL-nHAp scaffold and HOBs resulted in the synergistic enhancement of osteogenic gene expression. Moreover, it was found that BCP/PCL-nHAp scaffolds induced early osteogenic differentiation of ASCs through integrin-α2 and an extracellular signal-regulated kinase (ERK) signaling pathway (Lu et al. 2012).

7.2.2 Composites of Bioactive Glass (BG) or Mesoporous Bioactive Glass (MBG) and Polymers

Bioactive glass (BG) is another kind of inorganic material widely used in bone tissue engineering. Mesoporous BG (MBG), as a new class of BG, was first developed in 2004. Compared to traditional BG, MBG possesses a more optimal surface area and a well-ordered nanochannel structure, whose *in vitro* bioactivity is far superior than that of nonmesoporous BG (Figure 7.4) (Wu, Zhang, Zhou, et al. 2011).

7.2.2.1 Particles

Poly(epsilon-caprolactone) (PCL)/bioactive glass (BG) nanocomposite was fabricated using BG nanofibers (BGNFs) and compared with an established composite fabricated using microscale BG particles. The BGNFs were generated using sol-gel precursors via the electrospinning process, chopped into short fibers, and then incorporated into the PCL organic matrix by dissolving them in a tetrahydrofuran solvent. The biological and mechanical properties of the PCL/BGNF composites were evaluated and compared with those of PCL/BG powder (BGP). *In vitro* cell tests using the MC3T3 cell line demonstrated the enhanced

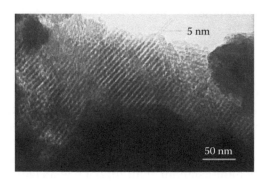

FIGURE 7.4
TEM image of the prepared MBG scaffolds. MBG scaffolds have a highly ordered mesopore-channel structure.

biocompatibility of the PCL/BGNF composite compared with the PCL/BGP composite. Furthermore, the PCL/BGNF composite showed a significantly higher level of bioactivity compared with the PCL/BGP composite. In addition, the results of the *in vivo* animal experiments using Sprague-Dawley albino rats revealed the good bone regeneration capability of the PCL/BGNF composite when implanted in a calvarial bone defect. In the result of the tensile test, the stiffness of the PCL/BG composite was further increased when the BGNFs were incorporated. These results indicate that the PCL/BGNF composite has greater bioactivity and mechanical stability when compared with the PCL/BG composite and great potential as a bone regenerative material (Jo et al. 2009).

Mohammadi et al. (2011) investigated the effect of Si and Fe on the surface properties of calcium-containing phosphate based glasses (PGs) in the system (50P$_2$O$_5$-40CaO-(10-x)SiO$_2$-xFe$_2$O$_3$, where x = 0, 5 and 10 mol%). Two PG formulations, 50P$_2$O$_5$-40CaO-10Fe$_2$O$_3$ (Fe10) and 50P$_2$O$_5$-40CaO-5Fe$_2$O$_3$-5SiO$_2$ (Fe5Si5), were melt drawn into fibers and randomly incorporated into poly(lactic acid) (PLA) produced by melt processing. In deionized water (DW), the dissolution rate of PLA-Fe5Si5 was significantly higher than that of PLA-Fe10. Dissolution of the glass fibers resulted in the formation of channels within the matrix. After phosphate buffered saline (PBS) aging, the reduction in mechanical properties was greater for PLA-Fe5Si5 compared to PLA-Fe10. MC3T3-E1 preosteoblasts seeded onto PG discs, PLA, and PLA-PGF composites were evaluated for up to 7 days indicating that the materials were generally cytocompatible. In addition, cell alignment along the PGF orientation was observed showing cell preference toward PGF (Mohammadi et al. 2011).

The *in vitro* osteogenic development of rat bone marrow mesenchymal stem cells (rBMSCs) upon different membrane substrates (pure PLA control, PLA-BG, and PLA-ZnBG) was investigated in terms of bone cell phenotype syntheses and mineralization. Oh et al. (2011) showed significantly stimulated production of ALP and osteocalcin at days 14 and 21 in the membranes containing BG and bioactive inorganic zinc-containing bioactive glass (ZnBG), with more in the samples containing ZnBG. The addition of ZnBG in poly(lactic acid) (PLA) allowed the rBMSCs to express a high level of bone sialoprotein as confirmed by immunostaining. Cellular mineralization of the secreted extracellular matrix showed a significantly higher Ca level on the BG- and ZnBG-added membrane than on the PLA, and more so in the ZnBG-added one (Oh et al. 2011).

As the amount of bioactive nanofiber increased (from 5% to 25%), the *in vitro* bioactivity of the nanocomposite was improved. The osteoblast responses to the nanocomposites (compositions with 5% and 25% nanofiber) were assessed in terms of cell proliferation, differentiation, and mineralization. Osteoblasts attached and grew well on the nanocomposites and secreted collagen protein at initial culturing periods. The differentiation of cells, as assessed by the expression of ALP, was significantly improved on the nanocomposites as compared to those on pure PLA. Moreover, the mineralized product by the cells was observed to be significantly higher on the nanocomposites with respect to pure PLA. The newly developed nanocomposite constituted of bioactive nanofiber and degradable polymer is considered as a promising bone regeneration matrix with its excellent bioactivity and osteoblast response (Kim et al. 2008).

7.2.2.2 Microspheres

Lei et al. (2012) examined the utility of sol-gel-derived bioactive glass microspheres (BGMs) as a reinforcement to improve the mechanical properties and biological performance of poly(ε-caprolactone) (PCL) polymer. All of the PCL-BGMs composites produced, with a variety of BGM contents (10, 20, and 30 wt%), showed a uniform distribution of the BGMs in the PCL matrix, particularly owing to their spherical shape and small size. This led to a considerable increase in the elastic modulus from 93±12 MPa to 635±179 MPa with increasing BGMs content from 0 to 30 wt%. Furthermore, the addition of the BGMs to the PCL polymer significantly increased the hydrophilicity of the PCL-BGMs composites, leading to a higher water absorption and degradation rate. The PCL-BGMs composite with a BGMs content of 30 wt% showed vigorous growth of apatite crystals with a high aspect ratio on its surface after soaking in the simulated body fluid for 7 days, resulting in the creation of a porous carbonate hydroxyapatite layer (Lei et al. 2012).

7.2.2.3 Scaffolds

Wu, Zhang, Zhu, et al. (2010) investigated the effects that blending silk incorporated into MBG on the physiochemical, drug delivery, and biological properties of MBG scaffolds. The results showed that the uniformity and continuity of the pore network of MBG/silk composites improved, and high porosity (94%) and large pore size (200–400 mm) maintained. The mechanical strength, mechanical stability, and control of burst release of DEX in silk-modified MBG scaffolds reached a great increase. Silk modification also appeared to provide a better environment for BMSC attachment, spreading, proliferation, and osteogenic differentiation on MBG/silk composites (Figure 7.5) (Wu, Zhang, Zhu, et al. 2010).

Wu, Zhang, Zhou, et al. (2011) used a freeze-drying method to incorporate MBG into silk scaffolds in order to improve their osteoconductivity and then to compare the effect of MBG and BG on the *in vivo* osteogenesis of silk scaffolds. The scaffolds were implanted into calvarial defects in SCID mice. The results showed that MBG/silk scaffolds have better physiochemical properties (mechanical strength, *in vitro* apatite mineralization, Si ion release, and pH stability) compared to BG/silk scaffolds. MBG and BG both improved the *in vivo* osteogenesis of silk scaffolds. lCT and HE analyses showed that

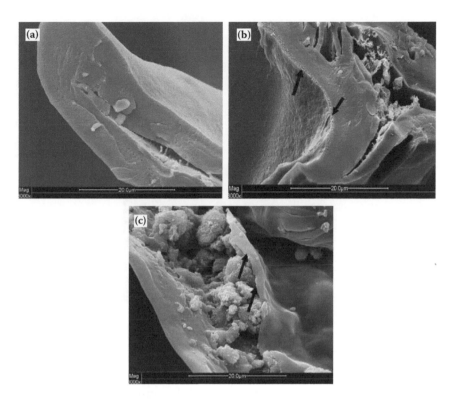

FIGURE 7.5
The cross section of pore walls for (a) 2.5 silk/MBG, (b) 5.0 silk/MBG, and (c) silk form a thin layer on the surface of pore walls.

MBG/silk scaffolds induced a slightly higher rate of new bone formation in the defects than did BG/silk scaffolds and immunohistochemical analysis showed greater synthesis of type I collagen in MBG/silk scaffolds compared to BG/silk scaffolds (Figure 7.6) (Wu, Zhang, Zhou, et al. 2011).

Georgiou et al. (2007) applied phosphate glass (PG) of the composition 0.46(CaO)-u0.04(Na(2)O)-0.5(P(2)O(5)) as filler in poly-L-lactic acid (PLA) foams to form degradable scaffolds for bone tissue engineering. BG share some properties with other bioactive inorganic material such as HAp and β-TCP, in which it can bond with host bone tissue, but BG, on the whole, have better bioactivity and degradation properties (Wu, Zhang, et al. 2010). Two different composite types were manufactured that contained either S2-high content silica S-BG, or A2-high content lime S-BG. The composites were evaluated in the form of sheets and 3D scaffolds. Sheets containing 12, 21, and 33 vol% of each bioactive glass were characterized for mechanical properties, wettability, hydrolytic degradation, and surface bioactivity. Sheets containing A2 S-BG rapidly formed a HAp layer after incubation in simulated body fluid.

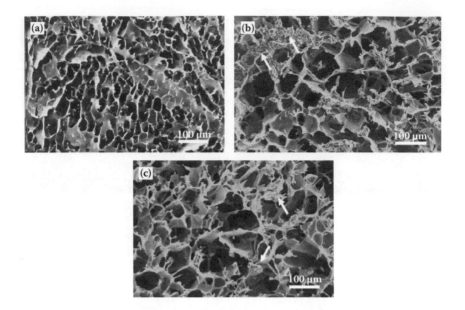

FIGURE 7.6
Surface morphology of (a) porous silk, (b) MBG/silk, and (c) BG/silk scaffolds. The arrows indicate MBG or BG particles in silk scaffolds.

Sheets and 3D scaffolds were evaluated for their ability to support growth of BMSCs and MG-63 cells, respectively. Cells were grown in nondifferentiating, osteogenic, or osteoclast-inducing conditions. Osteogenesis was induced with either recombinant human BMP-2 or dexamethasone, and osteoclast formation with M-CSF. BMC viability was lower at higher S-BG content, though specific ALP/cell was significantly higher on PLGA/A2-33 composites. Composites containing S2 S-BG enhanced the calcification of extracellular matrix by BMC, whereas incorporation of A2 S-BG in the composites promoted osteoclast formation from BMC. MG-63 osteoblast-like cells seeded in porous scaffolds containing S2 maintained viability and secreted collagen and calcium throughout the scaffolds. It is suggested that these sol-gel-derived bioactive glass-PLGA composites may prove excellent potential orthopedic and dental biomaterials supporting bone formation and remodeling (Pamula et al. 2011).

Jiang et al. (2010) prepared the scaffold based on agarose hydrogel or poly-lactide-co-glycolide (PLGA) and 45S5 bioactive glass (BG). It was observed that the stratified scaffold supported the region-specific coculture of chondrocytes and osteoblasts, which can lead to the production of three distinct yet continuous regions of cartilage, calcified cartilage, and bonelike matrices. Moreover, higher cell density enhanced chondrogenesis and improved graft mechanical property over time. The PLGA-BG phase promoted chondrocyte mineralization potential and is required for the formation of a calcified interface and bone regions on the osteochondral graft (Jiang et al. 2010).

7.2.3 Composites of Silicate Bioceramics and Polymers

Ceramic/polymer composites have been considered third-generation orthopedic biomaterials due to their ability to closely match properties (such as surface, chemistry, biological, and mechanical) of natural bone. Mixing bioactive ceramic powders with polymers is an effective method for generating bioactivity to the polymer-matrix composites, but it is necessary to incorporate up to 40 vol% of bioactive ceramic powder. However, such a high mixing ratio offsets the advantages of the flexibility and formability of polymer matrix and it would be highly advantageous to lower the mixing ratio.

7.2.3.1 Particles

Liu and Webster (2010) demonstrated that well-dispersed ceramic nanoparticles (titania or HAp) in PLGA improved mechanical properties compared with agglomerated ceramic nanoparticles even though the weight percentage of the ceramics was the same. Specifically, well-dispersed nanoceramics in PLGA enhanced the tensile modulus, tensile strength at yield, ultimate tensile strength, and compressive modulus compared with the more agglomerated nanoceramics in PLGA (Liu and Webster 2010). Since surface loading of ceramic powders in the polymer is thought to be an effective way of reducing the mixing ratio of the ceramic powder while maintaining bioactive activity, $CaSiO_3$/polylactic acid (PLA) composites were prepared by three methods: (1) casting, (2) spin coating, and (3) hot pressing. The bioactivity of these samples was investigated *in vitro* using simulated body fluid. Apatite formation was not observed in the samples prepared by method 1 but some apatite formation was achieved by mixing polyethylene glycol (PEG) with the PLA, producing a porous polymer matrix. In method 2, apatite was clearly observed after soaking for 7 days. Enhanced apatite formation was observed in method 3, the thickness of the resulting apatite layers is about 20 microm after soaking for 14 days. Since the amount of $CaSiO_3$ powders used in these samples was only ≤0.4 vol%, it is concluded that this preparation method is very effective in generating bioactivity in polymer-matrix composites by loading with only very small amounts of silicate ceramic powder (Okada et al. 2007).

7.2.3.2 Microspheres

CaSiO 3 (CS) has been proposed as a new class of material suitable for bone tissue repair due to its excellent bioactivity. Wu, Zhang, Fan, et al. (2011) incorporated CS into PLGA microspheres to investigate how the phase structure (amorphous and crystal) of CS influences the *in vitro* and *in vivo* bioactivity of the composite microspheres. The results showed that the incorporation of both amorphous-CS and crystal-CS enhanced the *in vitro* and *in vivo* bioactivity of PLGA microspheres. Crystal-CS/PLGA microspheres improved

FIGURE 7.7
SEM morphology for CaSiO$_3$ powders (a) before and (b) after calcining at 800°C.

better in vitro BMSC viability and bone-formation ability *in vivo* compared to amorphous-CS/PLGA microspheres. The study indicated that controlling the phase structure of CS is a promising method to modulate the bioactivity of polymer microsphere systems for potential bone tissue regeneration (Figure 7.7 and Figure 7.8) (Wu, Zhang, Fan, et al. 2011).

Microspheres have received significant attention as an injectable material for bone tissue regeneration. Compared with the traditional block scaffolds, the main advantage of this approach is that minute microspheres can be combined with a drug vehicle and be administered by injection, opening up the possibility of filling defects of various shapes and sizes through minimally invasive surgery (Wu, Zhang, Fan, et al. 2011). Nair et al. (2009) reported platelet-rich plasma (PRP) or fibrin glue (FG) in combination with

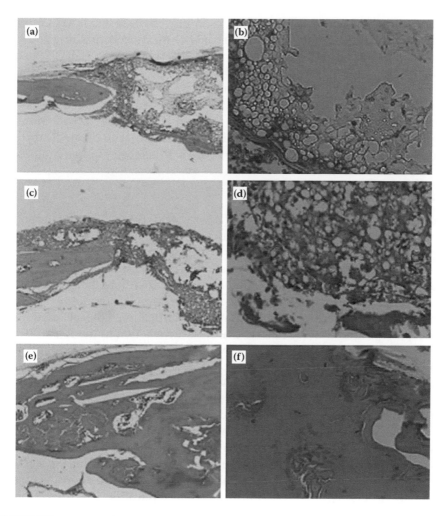

FIGURE 7.8
(**See color insert.**) The *in vivo* bone formation was assessed by (a) H&E for (b) PLGA micro-spheres, (c, d) for amorphous CaSiO₃/PLGA microspheres, and (e, f) for crystal CaSiO₃/PLGA microspheres. Panels (a), (c), and (e) are low magnification images by 4 times; (b), (d), and (f) are higher magnification images by 40 times.

bioactive ceramics could significantly enhance the functional activity of cells. They separated PRP and FG on bioactive ceramics like HAp and silica-coated HAp (HApSi), on which gBMSCs were seeded and induced to differentiate into the osteogenic lineage for 28 days. The results of this study have depicted that FG-coated ceramics were better than PRP-coated and bare matrices. Among all, the excellent performance was shown by FG coated HApSi, which may be attributed to the communal action of the stimulus emanated by Si in HApSi and the temporary extracellular matrix provided by FG over HApSi (Nair et al. 2009). Toskas et al. (2011) showed a ceramic silica (SiO(2)) hybrid nanofibers based on polyethylene oxide (PEO) and a new solution of modified sol-gel particles of mixture containing tetraethoxysilane (TEOS) and 3-glycidyloxypropyltriethoxysilane (GPTEOS) in a weight ratio of 3:1. Adding high-molecular-weight PEO into the silica sol is found to enhance the formation of the silica nanofibers and reduces the water-soluble carrying polymer to 1.2 wt%. These hybrid silica nanofibers possess unique collective properties that can have a great impact either in high-temperature reinforced materials and filtration or in biomedical applications such as in dentistry and bone tissue engineering (Toskas et al. 2011).

7.2.3.3 Scaffolds

Bioresorbable scaffolds made of poly(L-lactic acid) (PLA) obtained by supercritical gas foaming were recently described as suitable for tissue engineering, portraying biocompatibility with primary osteoblasts *in vitro* and mechanical properties when reinforced with ceramics. The behavior of such constructs remained to be evaluated *in vivo* and therefore the present study was undertaken to compare different PLA/ceramic composite scaffolds obtained by supercritical gas foaming in a critical size defect craniotomy model in Sprague-Dawley rats. The host-tissue reaction to the implants was evaluated semiquantitatively and similar tendencies were noted for all graft substitutes: initially highly reactive but decreasing with time implanted. Complete bone bridging was observed 18 weeks after implantation with PLA/5 wt% beta-TCP (PLA/TCP) and PLA/5 wt% HAp (PLA/HAp) scaffolds as assessed by histology and radiography (Montjovent et al. 2007).

Cunningham et al. (2010) demonstrated the potential use of marine sponges as precursors in the production of ceramic-based tissue-engineered bone scaffolds. They chose three species of natural sponge—Dalmata Fina (*Spongia officinalis Linnaeus*, Adriatic Sea), Fina Silk (*Spongia zimocca*, Mediterranean) and elephant ear (*Spongia agaricina*, Caribbean)—as replication. The most promising of the ceramic tissue engineered bone scaffolds developed, Spongia agaricina replicas, demonstrated an overall porosity of 56% to 61% with 83% of the pores ranging between 100 and 500 microm (average pore size 349 microm) and an interconnectivity of 99.92% (Cunningham et al. 2010).

7.3 Gene Delivery

7.3.1 Calcium Phosphate (CaP)-Based Hybrid Materials for Gene Delivery

Of all the inorganic vectors, CaP materials as the elementary constituent of bone and teeth have been so far mostly used due to their inherent material properties such as easily modified particle size, loading capacity, affinity to nucleic acid, biodegradability, and biocompatibility (Zhang, Wu, et al. 2010; Bongio et al. 2011; Holmquist et al. 2011). CaP materials have shown increased transfection efficacy as compared to naked plasmid DNA transfection *in vitro* and *in vivo*. Due to the uncertainty in CaP-DNA complex synthesis acted by pH, temperature, and butter conditions, the idea of applying CaP-DNA coprecipitation with organic polyplexes such as PEG, PLG, PVA, chitosan, and liposome was proposed and brought out high transfection efficiency and low toxicity (Fu et al. 2005; Ramachandran et al. 2009; Choi and Murphy 2010; Kimura et al. 2011).

To confirm the hybrid materials and nucleic acid to be an integrated gene delivery system, many methods are enrolled. Transmission electron microscopic (TEM) observation demonstrated the hybrid structure participated with CaP (Zohra et al. 2009). Scanning electron microscopic observation and dynamic light scattering measurement indicated that HAps were obviously encapsulated in the fabricated PVA/HAp/DNA nanoparticles (Kimura et al. 2011). Differential scanning calorimetry (DSC) suggested the existing interaction among each component of DNA-calcium-phosphate (CaPi-pDNA) complexes and the polymeric matrices of PLGA (Tang et al. 2012). X-ray diffractometry (XRD) further proved the conclusion that the CaPi-pDNA was in weak crystallization form inside the nanoparticles.

Several studies have focused on the factors to enhance the transfection efficiency. The intrinsic properties of the CaP mineral coating and the surrounding solution conditions are supposed to be the pacing factors in the release of plasmid DNA (Choi and Murphy 2010). The effect of the molecular weight, composition of organic phase, initial concentration of calcium phosphate, and Ca/P ratio were traversed based on a CaPi-pDNA-PLGA gene delivery system (Tang et al. 2012). The researcher demonstrated the optimal formulation to be spherical in shape, 207 ± 5 nm in size, and $95.7\%\pm0.8\%$ in entrapment efficiency. In addition, in another control study, apparently higher gravitational force rooted in the applying inorganic carbonate apatite enhanced initial steps of cellular contact and internalization of mRNA compared to separate cationic liposome (Zohra et al. 2009).

Crucially, the combined application of organics and CaP made up weaknesses mutually. To prevent the fast crystallization of CaP, which is the main factor for the transfection efficiency, Kakizawa et al. (2004) introduced a novel self-assembly of poly(ethylene glycol)-block-poly(asparticacid) block copolymers (PEG-P(Asp)) with CaP hybrid nanoparticles. Nevertheless, the application of chrloquin functioning as an endosomal-escape promoter has been restricted. In this regard, a modified strategy was brought out with poly(methacrylic acid) (PMA) in use revealing excellent colloidal stability and serum tolerability attribute to the steric stabilization effect of the PEG palisade (Kakizawa et al. 2006). To promote cell adhesion onto the hybrid layer, fibronectin was fabricated with supersaturated calcium phosphate solution supplemented with BMP-2 DNA and in this way enhanced gene transfer of the adhering cells (Wang, Oyane, et al. 2011). In contrast, nano-hydroxyapatite acted as a gene-activated matrix (GAM) for BMP2 delivery conjugated with bioactive and biodegradable collagen (Curtin et al. 2012). To decrease the toxicity of polymeric gene carriers, strategies of PEGylation and complexes formed via nonelectrostatic interactions, and such hydrogen bonding was proposed to combine medication with HAp using high hydro-static pressurization technology (Kimura et al. 2011).

On the strength of these mechanism studies, CaP-based hybrid materials are qualified vectors *in vitro* and *in vivo*. Various cell lines such as human embry-onic kidney 293 cell (Tang et al. 2012), C3H10T1/2 cells (Luong et al. 2009), human umbilical vein endothelial cell (Zohra et al. 2009), and COS-7 cells (Kimura et al. 2011) have been coverage initiated. Also, ectopic bone formation (Wang, Oyane, et al. 2011) and stem cell-mediated bone formation (Curtin et al. 2012) exhibited a promising gene delivery system for bone regeneration.

7.3.2 Hybrid Materials Integrated with Gold Nanoparticles (GNPs) for Gene Delivery

Gold nanoparticles (GNPs) have gained particular interest for widespread use as delivery platforms because of their advantageous characteristics that are easily surface functionalized with chemical and biological molecules, photothermal properties, and apparently low toxicity. One of the most promising applications is for gene delivery and transfection as a functioned vector. Moreover, modified with organic molecules containing thiols, phos-phates, and amines make their application more efficient.

Various organic materials including peptides, cationic polymers, and cat-ionic lipids have been extensively investigated as gene delivery vehicles. Among them, PEI has been widely exploited to form polyelectrolyte com-plexes with nucleic acid, and its excellent performance is becoming the standard for polymer-mediated gene delivery by reducing enzymatic deg-radation, enhancing cellular uptake, and endosome escape. Song et al. (2010) have conjugated PEI-capped AuNPs to siRNA targeting endogenous cell-cycle kinase displaying significant gene silence and enhanced cell apoptosis. The influence of particle size on cell transfection was investigated by two sets of PEI-coated Au NPs having particle size about 6 nm (<10 nm Au-PEI NPs) or 70 nm (<100 nm Au-PEI NPs), respectively (Cebrián et al. 2011). The smaller Au-PEI NPs led to higher efficiency owing to their lower agglom-eration state inside cells and quicker endosomal escape of DNA. However, 4 and 100 nm AuNPs showed similar biological effects *in vivo* by applying PEG-coated AuNPs as vector (Cho et al. 2009). In order to offer support for potential further clinical application, PEI-conjugated gold nanoparticles were applied to corneal tissues exhibiting appreciable gold uptake and mod-erate toxicity (Sharma et al. 2011).

Target-specific hybrid material is regarded as one of the most important platforms for the development of gene delivery. Hyaluronic acid (HAp) as a delivery carrier contributed to the highly efficient target-specific integrated with the liver tissues with HAp receptors like cluster determinant 44 (CD44) or hyaluronan receptor for endocytosis (HARE). Cysteamine modified gold nanoparticles (AuCM)/siRNA/PEI/HAp complex was successfully developed using a layer-by-layer method showing an excellent target-specific gene silencing effciency of ca. 70% in the presence of 50 vol% serum (Lee et al. 2011). High density lipoprotein (HDL) can target cancer cells that overexpress HDL receptors for dependence upon cholesterol delivery to maintain membrane biosynthesis and integrity (Ryan 2010). Also, biomimetic spherical HDL has been synthesized to mimic the size, shape, and surface chemistry of natural LDL. To take full advantage of this feature, HDL AuNPs successfully conjugated with antisense cholesterylated DNA regulated target gene expression (McMahon et al. 2011).

To enhance the uptake by anionic charge cell membrane, the conventional systems that were modified to be a positive surface charge might induce nonspecific binding with blood serum components resulting in aggregation and embolism in the body (Park et al. 2010). In contrast, the gold nanoparticles coated with PEI, PEG, and hyaluronic acid all showed negative zeta potential and the cell internalized efficiently in spite of the electrostatic repulsion that occurred between the surface of the cell and the hybrid surface (Kawano et al. 2006; Cebrián et al. 2011; Lee et al. 2011).

Meanwhile, the methods for fabrication of hybrid vectors were improved to avoid the complex surface functionalization, heterogeneous mixture, and potential toxicity. The idea of one-pot synthesized polypeptide-conjugated AuNPs was brought out by using positively charged polypeptides serving not only as capping agents but also as reductants subverting the classic ligand exchange strategy (Yan et al. 2012). Moreover, the instability of gold nanoparticles after centrifuged to remove cetyltrimethyl ammonium bromide limits its diverse biological applications. A delivery system by depositing gold nanoparticle surface with alternating layers of anionic and cationic polyelectrolytes was developed and demonstrated a stable and admirable photothermal response (Huang et al. 2010).

7.3.3 Silicate-Based Hybrid Material for Gene Delivery

Silicate nanoparticle has been considered a promising material due to the variety of available chemical and physical modifications. It can be tailored with a variety of modifiers by customizing the structures and physicochemical properties to achieve suitable particle size, stability, and high-loading capacity providing an efficient and safety carrier for gene delivery. Roy (2005) introduced an organically modified silica nanoparticle as DNA carriers, providing a promising direction for targeted therapy with enhanced efficacy as well as for real-time monitoring of drug action. Due to positively charged cationic-amino groups that were surface functionalized, the nanoparticle bound to DNA efficiently with individual fluorescence cross-linked.

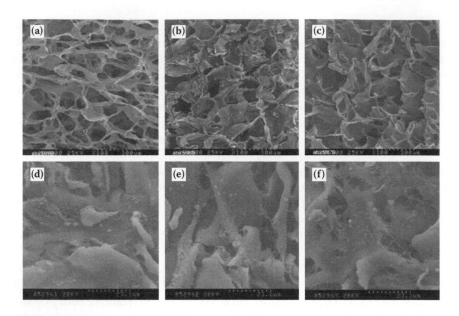

FIGURE 7.9
SEM of scaffold surface. (a) Pure chitosan/collagen scaffold, (b) the scaffolds with pEGFP-TGF-β1, (c) the scaffolds with AdTgf-β1, (d) HPLCs on pure chitosan/collagen scaffold after 2 days culture *in vitro*, (e) HPLCs on the scaffolds with pEGFP-TGF-β1 after 2 days culture *in vitro*, and (f) HPLCs on the scaffolds with AdTgf-β1 after 2 days culture *in vitro*.

The surface functionalization of silicate nanoparticle has been popular with an integrated organic molecule to expose amino groups. A kind of new self-assembled amino-modified silica nanoparticle was established with a diameter of 20 to 30 nm and positive surface charges of +11.3 mV (Xiao et al. 2009). Xiao et al. (2009) found that amino-modified silica nanoparticles could tightly bind with plasmid DNA, and the resulting nanoconjugate could transfer DNA to cells efficiently *in vitro* and *in vivo*, indicating the potential of gene therapy for hepatocellular carcinoma. The conjugated organics have also been modified to adapt to the strict environmental requirement. Chitosan is an N-deacetylation product of chitin, and it has been proved to be biologically renewable, biodegradable, biocompatible, nonantigenic, nontoxic, and biofunctional (Figure 7.9) (Zhang et al. 2006). Concerning its solubility limitations in a fluid environment, quaternized chitosan hybrid with rectorite exhibited better transfection efficiency (Wang et al. 2008).

For pursuing higher efficiency, the targeted gene delivery was explored tentatively. It is well known that mannose receptors are abundantly expressed on antigen presenting cells (APCs) such as macrophages (Singodia et al. 2012). In this regard, mannosylated polyethylenimine coupled mesoporous silica nanoparticles targeted macrophage cells with mannose receptors selectively and enhanced transfection efficiency (Park et al. 2008). Similarly, peptide-targeted mesoporous silica nanoparticle was fabricated to delivery siRNA (Ashley et al. 2012). In this study, the protocells conjugated with a peptide that specifically recognizes hepatocellular carcinomas gained a 106-fold improvement in efficacy compared to corresponding liposomes.

The cell internalization is the key step for gene delivery. It contains two kinds of process receptor-mediated or non-receptor-mediated path with the former more powerful. The particle size, shape, and surface charge are the important parameters for controlling the cell internalization. Preliminary studies indicated that the SiO_2 particles with a diameter of 200 nm were taken up by HepG2 cells with a faster rate than that of other sizes. As an effective receptor-mediated molecule, the Tat peptide was taken up with SiO_2 submicron particles demonstrating a significant fast rate of cell uptake (Mao et al. 2010). Moreover, it maintained effectiveness in the low temperature of 4°C.

Interestingly, PEGylated mesoporous silica nanoparticles made codelivery of chloroquine and the nucleic acids with a significantly increased transfection and silencing activity, which in turn pointed out the orientation for the combination of drug–nucleic acid therapies (Bhattarai et al. 2010).

7.3.4 Quantum Dots-Based Hybrid Materials for Gene Delivery

Quantum dots (QDs) are inorganic nanoparticles that exhibit superior optical properties such as stability toward photobleaching, controllable emission bands, broad absorption spectra, and high quantum yields, which in turn were widely used in biolabeling and bioimaging (Medintz et al. 2005; Kim et al. 2010). In addition, they have aroused interest for gene delivery owing to their good biocompatibility and monodispersity (Medintz et al. 2005).

Surface functionalization was achieved to hit better gene delivery. To condense plasmid DNA into nanocomplexes, poly(2-(dimethylami-no)ethyl methacrylate), a polycation, was used to modify ZnO quantum dot (Zhang and Liu 2010). To reduce the quantum yields of the QDs drop and cytotoxicity following the conventional ligand-exchange reactions, QD nanoparticles were coated with b-cyclodextrin and then coupled to amino acids (Zhao et al. 2011). Commonly used PEGylated nanoparticles not only reduce their non-specific cellular uptake, which was mostly caused by surface protein adsorption in the serum, but also showed significantly reduced extent of cellular uptake to targeted cells (Malek et al. 2008; Jeong et al. 2009). The shielding/deshielding of cell penetrating peptides (CPP) was regulated by morphological transform of poly (N-iso-propylacrylamide) due to the lower critical solution temperature. Based on this principle, the complex of quantum dots modified with CPP exhibit an "on-demand" cellular uptake behavior (Kim et al. 2010).

7.3.5 Iron-Based Hybrid Materials for Gene Delivery

Iron-based nanoparticle as a kind of magnetic nanoparticles has gained considerable attention due to their promising efficiency, nontoxicity, biocompatibility. and unique superparamagnetic characteristic as a nonviral vector (Arsianti et al. 2010). However, its fast aggregate propensity and low colloidal stability needed modification to achieve better gene delivery (Chorny et al. 2007).

Core/shell Fe3O4@SiO2 nanoparticles modified with PAH was manufactured as a vector for EGFP plasmid DNA delivery demonstrating a fairly high expression level (Shi et al. 2011). Similarly, taking advantage of the synergistic effect, covalently conjugated PEI and superparamagnetic iron oxide nanoparticles was confirmed to be conducive to effective DNA binding and enhanced cellular uptake (Namgung et al. 2010). Despite the progress, such coated magnetic nanoparticles displayed only moderate transfection efficiency compared to other kinds of hybrid materials.

7.4 Conclusion

By introducing the three kinds of inorganic and organic composites materials for applications in tissue engineering and bone regeneration and gene delivery, we learn a lot about the latest development in bone repair. As the first generation materials applied in bone regeneration, CaP inorganic materials have the same chemical ingredients as the natural bone. But without cell and blood nurturing, CaP inorganic materials' physical and biological characteristics cannot satisfy the requirements of bone regeneration. For porous mesopore bioglass (MBG), proposed as a new class of bone regeneration materials, their apatite-formation and drug-delivery properties promises it a good development in bone regeneration. The highly ordered mesoporous arrangement of cavities permits the confinement of different drug molecules to be subsequently released, acting as controlled delivery systems. However, the material's inherent brittleness and high degradation and surface instability are major disadvantages, which compromise its mechanical strength and cytocompatibility as a biological scaffold. Actually, different composites have their own advantages, but we do not find the best composite materials in bone repair and regeneration. As to gene delivery, organics' fast degradation and weak mechanical strength, and inorganics' weak interactions between the inorganic vector and the ribonucleic acid molecules restrict their applications. Under this situation, bioactive organic and inorganic composite materials offer unique biological, electrical, and mechanical properties key to designing an excellent gene delivery system.

References

Arsianti, M., Lim, M., Marquis, C. P., et al. 2010. Assembly of polyethylenimine-based magnetic iron oxide vectors: Insights into gene delivery. *Langmuir* 26(10): 7314–7326.

Ashley, C. E., Carnes, E. C., Epler, K. E., et al. 2012. Delivery of small interfering RNA by peptide-targeted mesoporous silica nanoparticle-supported lipid bilayers. *ACS Nano* 6(3): 2174–2188.

Bhattarai, S. R., Muthuswamy, E., Wani, A., et al. 2010. Enhanced gene and siRNA delivery by polycation-modified mesoporous silica nanoparticles loaded with chloroquine. *Pharm Res* 27(12): 2556–2568.

Bhumiratana, S., Grayson, W. L., Castaneda, A., et al. 2011. Nucleation and growth of mineralized bone matrix on silk-hydroxyapatite composite scaffolds. *Biomaterials* 32(11): 2812–2820.

Boccaccini, A. R., and Blaker, J. J. 2005. Bioactive composite materials for tissue engineering scaffolds. *Expert Rev Med Devices* 2(3): 303–317.

Bongio, M., van den Beucken, J. J., Nejadnik, M. R., et al. 2011. Biomimetic modification of synthetic hydrogels by incorporation of adhesive peptides and calcium phosphate nanoparticles: *In vitro* evaluation of cell behavior. *Eur Cell Mater* 22: 359–376.

Cao, H., and Kuboyama, N. 2010. A biodegradable porous composite scaffold of PGA/beta-TCP for bone tissue engineering. *Bone* 46(2): 386–395.

Causa, F., Netti, P. A., Ambrosio, L., et al. 2006. Poly-epsilon-caprolactone/hydroxy-apatite composites for bone regeneration: *In vitro* characterization and human osteoblast response. *J Biomed Mater Res A* 76(1): 151–162.

Cebrián, V., Martín-Saavedra, F., Yagüe, C., et al. 2011. Size-dependent transfection efficiency of PEI-coated gold nanoparticles. *Acta Biomaterialia* 7(10): 3645–3655.

Chen, J., Yu, Q., Zhang, G., et al. 2012. Preparation and biocompatibility of nanohybrid scaffolds by in situ homogeneous formation of nano hydroxyapatite from biopolymer polyelectrolyte complex for bone repair applications. *Colloids Surf B Biointerfaces* 93: 100–107.

Cho, W.-S., Kim, S., Han, B. S., et al. 2009. Comparison of gene expression profiles in mice liver following intravenous injection of 4 and 100 nm-sized PEG-coated gold nanoparticles. *Toxicol Lett* 191(1): 96–102.

Choi, S., and Murphy, W. L. 2010. Sustained plasmid DNA release from dissolving mineral coatings. *Acta Biomaterialia* 6(9): 3426–3435.

Chorny, M., Polyak, B., Alferiev, I. S., et al. 2007. Magnetically driven plasmid DNA delivery with biodegradable polymeric nanoparticles. *FASEB J* 21(10): 2510–2519.

Chowdhury, E. H., and Akaike, T. 2005. Bio-functional inorganic materials: An attractive branch of gene-based nano-medicine delivery for 21st century. *Curr Gene Ther* 5(6): 669–676.

Compston, J., and Rosen, C. J. 1999. *Fast Facts: Osteoporosis*, 2nd ed. Oxford: Health Press Limited.

Cunningham, E., Dunne, N., Walker, G., et al. 2010. Hydroxyapatite bone substitutes developed via replication of natural marine sponges. *J Mater Sci Mater Med* 21(8): 2255–2261.

Curtin, C. M., Cunniffe, G. M., Lyons, F. G., et al. 2012. Innovative collagen nano-hydroxyapatite scaffolds offer a highly efficient non-viral gene delivery platform for stem cell-mediated bone formation. *Adv Mater* 24(6): 749–754.

Di Martino, A., Sittinger, M., and Risbud, M. V. 2005. Chitosan: A versatile biopolymer for orthopaedic tissue-engineering. *Biomaterials* 26(30): 5983–5990.

Fu, H., Hu, Y., McNelis, T., et al. 2005. A calcium phosphate-based gene delivery system. *J Biomed Mater Res A* 74(1): 40–48.

Gaharwar, A. K., Dammu, S. A., Canter, J. M., et al. 2011. Highly extensible, tough, and elastomeric nanocomposite hydrogels from poly(ethylene glycol) and hydroxyapatite nanoparticles. *Biomacromolecules* 12(5): 1641–1650.

Georgiou, G., Mathieu, L., Pioletti, D. P., et al. 2007. Polylactic acid-phosphate glass composite foams as scaffolds for bone tissue engineering. *J Biomed Mater Res B Appl Biomater* 80(2): 322–331.

Gómez-Romero, P., and Sanchez, C. 2004. *Functional Hybrid Materials*. Weinheim: Wiley-VCH.

Guarino, V., Causa, F., Taddei, P., et al. 2008. Polylactic acid fibre-reinforced polycaprolactone scaffolds for bone tissue engineering. *Biomaterials* 29(27): 3662–3670.

Haaparanta, A. M., Haimi, S., Ella, V., et al. 2010. Porous polylactide/beta-tricalcium phosphate composite scaffolds for tissue engineering applications. *J Tissue Eng Regen Med* 4(5): 366–373.

Haimi, S., Suuriniemi, N., Haaparanta, A. M., et al. 2009. Growth and osteogenic differentiation of adipose stem cells on PLA/bioactive glass and PLA/beta-TCP scaffolds. *Tissue Eng Part A* 15(7): 1473–1480.

Holmquist, P. C., Holmquist, G. P., and Summers, M. L. 2011. Comparing binding site information to binding affinity reveals that Crp/DNA complexes have several distinct binding conformers. *Nucleic Acids Res* 39(15): 6813–6824.

Huang, H. C., Barua, S., Kay, D. B., et al. 2010. Simultaneous enhancement of photothermal stability and gene delivery efficacy of gold nanorods using polyelectrolytes. *ACS Nano* 3(10): 2941–2952; erratum, 4(3): 1769–1770.

Huang, X., and Miao, X. 2007. Novel porous hydroxyapatite prepared by combining H2O2 foaming with PU sponge and modified with PLGA and bioactive glass. *J Biomater Appl* 21(4): 351–374.

Hwang, S. J., and Davis, M. E. 2001. Cationic polymers for gene delivery: Designs for overcoming barriers to systemic administration. *Curr Opin Mol Ther* 3(2): 183–191.

Jeong, J. H., Mok, H., Oh, Y. K., et al. 2009. siRNA conjugate delivery systems. *Bioconjug Chem* 20(1): 5–14.

Jevtic, M., Radulovic, A., Ignjatovic, N., et al. 2009. Controlled assembly of poly(D,L-lactide-co-glycolide)/hydroxyapatite core-shell nanospheres under ultrasonic irradiation. *Acta Biomater* 5(1): 208–218.

Jiang, J., Tang, A., Ateshian, G. A., et al. 2010. Bioactive stratified polymer ceramic-hydrogel scaffold for integrative osteochondral repair. *Ann Biomed Eng* 38(6): 2183–2196.

Jo, J. H., Lee, E. J., Shin, D. S., et al. 2009. *In vitro/in vivo* biocompatibility and mechanical properties of bioactive glass nanofiber and poly(epsilon-caprolactone) composite materials. *J Biomed Mater Res B Appl Biomater* 91(1): 213–220.

Kakizawa, Y., Furukawa, S., Ishii, A., et al. 2006. Organic–inorganic hybrid-nanocarrier of siRNA constructing through the self-assembly of calcium phosphate and PEG-based block aniomer. *J Controlled Release* 111(3): 368–370.

Kakizawa, Y., Furukawa, S., and Kataoka, K. 2004. Block copolymer-coated calcium phosphate nanoparticles sensing intracellular environment for oligodeoxynucleotide and siRNA delivery. *J Controlled Release* 97(2): 345–356.

Kamimura, K., Suda, T., Zhang, G., et al. 2011. Advances in gene delivery systems. *Pharma Med* 25(5): 293–306.

Kawano, T., Yamagata, M., Takahashi, H., et al. 2006. Stabilizing of plasmid DNA *in vivo* by PEG-modified cationic gold nanoparticles and the gene expression assisted with electrical pulses. *J Controlled Release* 111(3): 382–389.

Kesheng, W., Kovacs, G. L., and Wozny, M., et al. 2006. *Knowledge Enterprise: Intelligent Strategies in Product Design, Manufacturing, and Management.* New York, NY: Springer Science+Business Media.

Kim, C., Lee, Y., Kim, J. S., et al. 2010. Thermally triggered cellular uptake of quantum dots immobilized with poly(N-isopropylacrylamide) and cell penetrating peptide. *Langmuir* 26(18): 14965–14969.

Kim, H. W. 2007. Biomedical nanocomposites of hydroxyapatite/polycaprolactone obtained by surfactant mediation. *J Biomed Mater Res A* 83(1): 169–177.

Kim, H. W., Lee, H. H., and Chun, G. S. 2008. Bioactivity and osteoblast responses of novel biomedical nanocomposites of bioactive glass nanofiber filled poly(lactic acid). *J Biomed Mater Res A* 85(3): 651–663.

Kim, J., McBride, S., Tellis, B., et al. 2012. Rapid-prototyped PLGA/beta-TCP/hydroxyapatite nanocomposite scaffolds in a rabbit femoral defect model. *Biofabrication* 4(2): 025003.

Kim, T. N., Feng, Q. L., Kim, J. O., et al. 1998. Antimicrobial effects of metal ions (Ag+, Cu2+, Zn2+) in hydroxyapatite. *J Mater Sci Mater Med* 9(3): 129–134.

Kimura, T., Nibe, Y., Funamoto, S., et al. 2011. Preparation of a nanoscaled poly(vinyl alcohol)/hydroxyapatite/DNA complex using high hydrostatic pressure technology for *in vitro* and *in vivo* gene delivery. *J Drug Delivery* 2011: 1–8.

Kootstra, N. A., and Verma, I. M. 2003. Gene therapy with viral vectors. *Annu Rev Pharm Toxicol* 43: 413–439.

Kothapalli, C. R., Shaw, M. T., and Wei, M. 2005. Biodegradable HA-PLA 3-D porous scaffolds: Effect of nano-sized filler content on scaffold properties. *Acta Biomater* 1(6): 653–662.

Langer, R., and Vacanti, J. P. 1993. Tissue engineering. *Science* 260(5110): 920–926.

Lee, H. H., Yu, H. S., Jang, J. H., et al. 2008. Bioactivity improvement of poly(epsilon-caprolactone) membrane with the addition of nanofibrous bioactive glass. *Acta Biomater* 4(3): 622–629.

Lee, M. Y., Park, S. J., Park, K., et al. 2011. Target-specific gene silencing of layer-by-layer assembled gold-cysteamine/siRNA/PEI/HA nanocomplex. *ACS Nano* 5(8): 6138–6147.

Lei, B., Shin, K. H., Noh, D. Y., et al. 2012. Bioactive glass microspheres as reinforcement for improving the mechanical properties and biological performance of poly(epsilon-caprolactone) polymer for bone tissue regeneration. *J Biomed Mater Res B Appl Biomater* 100(4): 967–975.

Liu, H., and Webster, T. J. 2010. Mechanical properties of dispersed ceramic nanoparticles in polymer composites for orthopedic applications. *Int J Nanomedicine* 5: 299–313.

Liu, L., Guo, Y., Chen, X., et al. 2012. Three-dimensional dynamic culture of preosteoblasts seeded in HA-CS/Col/nHAP composite scaffolds and treated with alpha-ZAL. *Acta Biochim Biophys Sin* (Shanghai) 44(8): 669–677.

Lu, Z., Roohani-Esfahani, S. I., Wang, G., et al. 2012. Bone biomimetic microenvironment induces osteogenic differentiation of adipose tissue-derived mesenchymal stem cells. *Nanomedicine* 8(4): 507–515.

Luo, D., and Saltzman, W. M. 2000. Synthetic DNA delivery systems. *Nat Biotechnol* 18(1): 33–37.

Luong, L. N., McFalls, K. M., and Kohn, D. H. 2009. Gene delivery via DNA incorporation within a biomimetic apatite coating. *Biomaterials* 30(36): 6996–7004.

Luvizuto, E. R., Tangl, S., Zanoni, G., et al. 2011. The effect of BMP-2 on the osteoconductive properties of beta-tricalcium phosphate in rat calvaria defects. *Biomaterials* 32(15): 3855–3861.

Malek, A., Czubayko, F., and Aigner, A. 2008. PEG grafting of polyethylenimine (PEI) exerts different effects on DNA transfection and siRNA-induced gene targeting efficacy. *J Drug Target* 16(2): 124–139.

Mao, Z., Wan, L., Hu, L., et al. 2010. Tat peptide mediated cellular uptake of SiO2 submicron particles. *Colloids Surfaces B: Biointerfaces* 75(2): 432–440.

McCullen, S. D., Zhu, Y., Bernacki, S. H., et al. 2009. Electrospun composite poly(L-lactic acid)/tricalcium phosphate scaffolds induce proliferation and osteogenic differentiation of human adipose-derived stem cells. *Biomed Mater* 4(3): 035002.

McMahon, K. M., Mutharasan, R. K., Tripathy, S., et al. 2011. Biomimetic high density lipoprotein nanoparticles for nucleic acid delivery. *Nano Letters* 11(3): 1208–1214.

Medintz, I. L., Uyeda, H. T., Goldman, E. R., et al. 2005. Quantum dot bioconjugates for imaging, labelling and sensing. *Nat Mater* 4(6): 435–446.

Mohammadi, M. S., Ahmed, I., Muja, N., et al. 2011. Effect of phosphate-based glass fibre surface properties on thermally produced poly(lactic acid) matrix composites. *J Mater Sci Mater Med* 22(12): 2659–2672.

Montjovent, M. O., Mathieu, L., Schmoekel, H., et al. 2007. Repair of critical size defects in the rat cranium using ceramic-reinforced PLA scaffolds obtained by supercritical gas foaming. *J Biomed Mater Res A* 83(1): 41–51.

Nair, M. B., Varma, H. K., and John, A. 2009. Platelet-rich plasma and fibrin glue-coated bioactive ceramics enhance growth and differentiation of goat bone marrow-derived stem cells. *Tissue Eng Part A* 15(7): 1619–1631.

Namgung, R., Singha, K., Yu, M. K., et al. 2010. Hybrid superparamagnetic iron oxide nanoparticle-branched polyethylenimine magnetoplexes for gene transfection of vascular endothelial cells. *Biomaterials* 31(14): 4204–4213.

Oh, S. A., Won, J. E., and Kim, H. W. 2011. Composite membranes of poly(lactic acid) with zinc-added bioactive glass as a guiding matrix for osteogenic differentiation of bone marrow mesenchymal stem cells. *J Biomater Appl.* DOI: 10.1177/0885328211408944.

Okada, K., Hasegawa, F., Kameshima, Y., et al. 2007. Bioactivity of CaSiO3/poly-lactic acid (PLA) composites prepared by various surface loading methods of CaSiO3 powder. *J Mater Sci Mater Med* 18(8): 1605–1612.

Pamula, E., Kokoszka, J., Cholewa-Kowalska, K., et al. 2011. Degradation, bioactivity, and osteogenic potential of composites made of PLGA and two different sol-gel bioactive glasses. *Ann Biomed Eng* 39(8): 2114–2129.

Park, I. Y., Kim, I. Y., Yoo, M. K., et al. 2008. Mannosylated polyethylenimine coupled mesoporous silica nanoparticles for receptor-mediated gene delivery. *Int J Pharm* 359(1–2): 280–287.

Park, K., Lee, M. Y., Kim, K. S., et al. 2010. Target specific tumor treatment by VEGF siRNA complexed with reducible polyethyleneimine-hyaluronic acid conjugate. *Biomaterials* 31(19): 5258–5265.

Peng, Q., Jiang, F., Huang, P., et al. 2010. A novel porous bioceramics scaffold by accumulating hydroxyapatite spherules for large bone tissue engineering *in vivo*. I. Preparation and characterization of scaffold. *J Biomed Mater Res A* 93(3): 920–929.

Piskin, E., Dincer, S., and Turk, M. 2004. Gene delivery: Intelligent but just at the beginning. *J Biomater Sci Polym Ed* 15(9): 1181–1202.

Ramachandran, R., Paul, W., and Sharma, C. P. 2009. Synthesis and characterization of PEGylated calcium phosphate nanoparticles for oral insulin delivery. *J Biomed Mater Res B Appl Biomater* 88(1): 41–48.

Ratner, B. D., and Bryant, S. J. 2004. Biomaterials: Where we have been and where we are going. *Annu Rev Biomed Eng* 6: 41–75.

Reichert, J. C., Epari, D. R., Wullschleger, M. E., et al. 2012. Bone tissue engineering. Reconstruction of critical sized segmental bone defects in the ovine tibia. *Orthopade* 41(4): 280–287. (Article in German)

Roohani-Esfahani, S. I., Lu, Z. F., Li, J. J., et al. 2012. Effect of self-assembled nanofibrous silk/polycaprolactone layer on the osteoconductivity and mechanical properties of biphasic calcium phosphate scaffolds. *Acta Biomater* 8(1): 302–312.

Roohani-Esfahani, S. I., Nouri-Khorasani, S., Lu, Z., et al. 2010. The influence hydroxyapatite nanoparticle shape and size on the properties of biphasic calcium phosphate scaffolds coated with hydroxyapatite-PCL composites. *Biomaterials* 31(21): 5498–5509.

Roohani-Esfahani, S. I., Nouri-Khorasani, S., Lu, Z. F., et al. 2011. Effects of bioactive glass nanoparticles on the mechanical and biological behavior of composite coated scaffolds. *Acta Biomater* 7(3): 1307–1318.

Roth, C. M., and Sundaram, S. 2004. Engineering synthetic vectors for improved DNA delivery: Insights from intracellular pathways. *Annu Rev Biomed Eng* 6: 397–426.

Roy, I. 2005. Optical tracking of organically modified silica nanoparticles as DNA carriers: A nonviral, nanomedicine approach for gene delivery. *Proc Nat Acad Sci* 102(2): 279–284.

Ryan, R. O. 2010. Nanobiotechnology applications of reconstituted high density lipoprotein. *J Nanobiotechnology* 8: 28.

Scaduto, A. A., and Lieberman, J. R. 1999. Gene therapy for osteoinduction. *Orthop Clin North Am* 30(4): 625–633.

Schaffer, D. V., and Lauffenburger, D. A. 1998. Optimization of cell surface binding enhances efficiency and specificity of molecular conjugate gene delivery. *J Biol Chem* 273(43): 28004–28009.

Shane, E. 2010. Evolving data about subtrochanteric fractures and bisphosphonates. *N Engl J Med* 362(19): 1825–1827.

Sharma, A., Tandon, A., Tovey, J. C. K., et al. 2011. Polyethylenimine-conjugated gold nanoparticles: Gene transfer potential and low toxicity in the cornea. *Nanomedicine* 7(4): 505–513.

Shi, M., Liu, Y., Xu, M., et al. 2011. Core/shell Fe3O4@SiO2 nanoparticles modified with PAH as a vector for EGFP plasmid DNA delivery into HeLa cells. *Macromol Biosci* 11(11): 1563–1569.

Singodia, D., Verma, A., Verma, R. K., et al. 2012. Investigations into an alternate approach to target mannose receptors on macrophages using 4-sulfated N-acetyl galactosamine more efficiently in comparison with mannose-decorated liposomes: An application in drug delivery. *Nanomedicine* 8(4): 468–477.

Sokolova, V., and Epple, M. 2008. Inorganic nanoparticles as carriers of nucleic acids into cells. *Angew Chem Int Ed Engl* 47(8): 1382–1395.

Son, J. S., Kim, S. G., Oh, J. S., et al. 2011. Hydroxyapatite/polylactide biphasic combination scaffold loaded with dexamethasone for bone regeneration. *J Biomed Mater Res A* 99(4): 638–647.

Song, W.-J., Du, J.-Z., Sun, T.-M., et al. 2010. Gold nanoparticles capped with polyethyleneimine for enhanced siRNA delivery. *Small* 6(2): 239–246.

Sundaram, S., Trivedi, R., Durairaj, C., et al. 2009. Targeted drug and gene delivery systems for lung cancer therapy. *Clin Cancer Res* 15(23): 7299–7308.

Tang, J., Chen, J.-Y., Liu, J., et al. 2012. Calcium phosphate embedded PLGA nanoparticles: A promising gene delivery vector with high gene loading and transfection efficiency. *Int J Pharm* 431(1–2): 210–221.

Toskas, G., Cherif, C., Hund, R. D., et al. 2011. Inorganic/organic (SiO(2))/PEO hybrid electrospun nanofibers produced from a modified sol and their surface modification possibilities. *ACS Appl Mater Interfaces* 3(9): 3673–3681.

Vallet-Regi, M., Colilla, M., and Gonzalez, B. 2011. Medical applications of organic–inorganic hybrid materials within the field of silica-based bioceramics. *Chem Soc Rev* 40(2): 596–607.

Van der Pol, U., Mathieu, L., Zeiter, S., et al. 2010. Augmentation of bone defect healing using a new biocomposite scaffold: An *in vivo* study in sheep. *Acta Biomater* 6(9): 3755–3762.

Wagner, D. E., and Bhaduri, S. B. 2012. Progress and outlook of inorganic nanoparticles for delivery of nucleic acid sequences related to orthopedic pathologies: A review. *Tissue Eng B Rev* 18(1): 1–14.

Wang, T., Chow, L. C., Frukhtbeyn, S. A., et al. 2011. Improve the strength of PLA/HA composite through the use of surface initiated polymerization and phosphonic acid coupling agent. *J Res Natl Inst Stand Technol* 116(5): 785–796.

Wang, X., Oyane, A., Tsurushima, H., et al. 2011. BMP-2 and ALP gene expression induced by a BMP-2 gene–fibronectin–apatite composite layer. *Biomed Mater* 6(4): 045004.

Wang, X., Pei, X., Du, Y., et al. 2008. Quaternized chitosan/rectorite intercalative materials for a gene delivery system. *Nanotechnology* 19(37): 375102.

Wang, J., Zhou, W., Hu, W., Zhou, L., Wang, S., and Zhang, S. 2011. Collagen/silk fibroin bi-template induced biomimetic bone-like substitutes. *J Biomed Mater Res A* 99(3): 327–334.

Wu, C., Zhang, Y., Fan, W., et al. 2011. CaSiO(3) microstructure modulating the *in vitro* and *in vivo* bioactivity of poly(lactide-co-glycolide) microspheres. *J Biomed Mater Res A* 98(1): 122–131.

Wu, C., Zhang, Y., Zhou, Y., et al. 2011. A comparative study of mesoporous glass/silk and non-mesoporous glass/silk scaffolds: Physiochemistry and *in vivo* osteogenesis. *Acta Biomater* 7(5): 2229–2236.

Wu, C., Zhang, Y., Zhu, Y., et al. 2010. Structure-property relationships of silk-modified mesoporous bioglass scaffolds. *Biomaterials* 31(13): 3429–3438.

Wu, G. J., Zhou, L. Z., Wang, K. W., et al. 2010. Hydroxylapatite nanorods: An efficient and promising carrier for gene transfection. *J Colloid Interface Sci* 345(2): 427–432.

Xiao, X., He, Q., and Huang, K. 2009. Novel amino-modified silica nanoparticles as efficient vector for hepatocellular carcinoma gene therapy. *Med Oncol* 27(4): 1200–1207.

Xu, H., Su, J., Sun, J., et al. 2012. Preparation and characterization of new nano-composite scaffolds loaded with vascular stents. *Int J Mol Sci* 13(3): 3366–3381.

Yan, X., Blacklock, J., Li, J., et al. 2012. One-pot synthesis of polypeptide-gold nano-conjugates for *in vitro* gene transfection. *ACS Nano* 6(1): 111–117.

Yang, M. C., Wang, S. S., Chou, N. K., et al. 2009. The cardiomyogenic differentiation of rat mesenchymal stem cells on silk fibroin-polysaccharide cardiac patches *in vitro*. *Biomaterials* 30(22): 3757–3765.

Yanoso-Scholl, L., Jacobson, J. A., Bradica, G., et al. 2010. Evaluation of dense poly-lactic acid/beta-tricalcium phosphate scaffolds for bone tissue engineering. *J Biomed Mater Res A* 95(3): 717–726.

Yeo, M., Lee, H., and Kim, G. 2011. Three-dimensional hierarchical composite scaf-folds consisting of polycaprolactone, beta-tricalcium phosphate, and collagen nanofibers: Fabrication, physical properties, and *in vitro* cell activity for bone tissue regeneration. *Biomacromolecules* 12(2): 502–510.

Zhang, P., and Liu, W. 2010. ZnO QD@PMAA-co-PDMAEMA nonviral vector for plasmid DNA delivery and bioimaging. *Biomaterials* 31(11): 3087–3094.

Zhang, X., and Godbey, W. T. 2006. Viral vectors for gene delivery in tissue engineer-ing. *Adv Drug Deliv Rev* 58(4): 515–534.

Zhang, Y., Cheng, X., Wang, J., et al. 2006. Novel chitosan/collagen scaffold contain-ing transforming growth factor-beta1 DNA for periodontal htissue engineering. *Biochem Biophys Res Commun* 344(1): 362–369.

Zhang, Y., Shi, B., Li, C., et al. 2009. The synergetic bone-forming effects of combina-tions of growth factors expressed by adenovirus vectors on chitosan/collagen scaffolds. *J Controlled Release* 136(3): 172–178.

Zhang, Y., Wu, C., Friis, T., et al. 2010. The osteogenic properties of CaP/silk compos-ite scaffolds. *Biomaterials* 31(10): 2848–2856.

Zhao, M.-X., Li, J.-M., Du, L., et al. 2011. Targeted cellular uptake and siRNA silenc-ing by quantum-dot nanoparticles coated with β-cyclodextrin coupled to amino acids. *Chemistry* 17(18): 5171–5179.

Zohra, F. T., Chowdhury, E. H., and Akaike, T. 2009. High performance mRNA trans-fection through carbonate apatite–cationic liposome conjugates. *Biomaterials* 30(23–24): 4006–4013.

Index

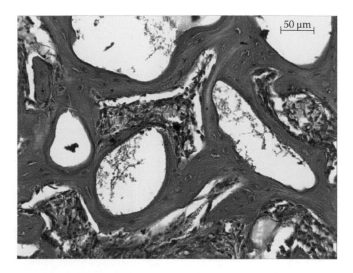

FIGURE 1.9
New bone formation around the MBG particles after implanted into the defects of rat femur for 8 weeks (red area: new bone; white area: MBG particles).

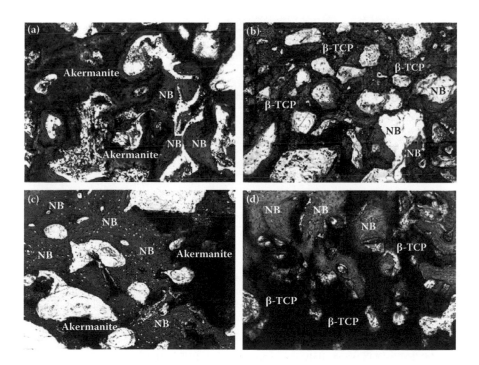

FIGURE 2.8
High magnification images of new bone formation and material degradation of (a, c) aker-manite and (b, c) β-TCP implants after (a, b) 8 and (c, d) 16 weeks (Van Gieson's picrofuchsin staining of transverse section; NB: new bone). Red color indicates newly formed bone. Original magnification: 100×.

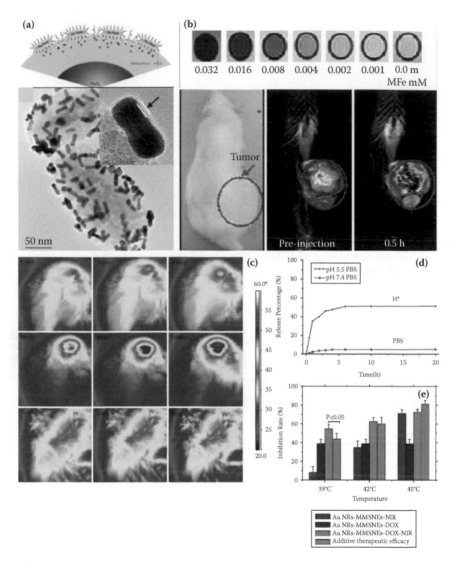

FIGURE 3.12
(a) Schematic microscopic structure and TEM image of Au NRs-MMSNEs. (b) T_2 phantom images of Au NRs-MMSNEs at different Fe concentrations and *in vivo* MRI of a mouse before and after intratumor injection of Au NRs-MMSNEs. (c) Infrared thermal imaging under the photothermal heating by 808-nm laser irradiation for different time periods in Au NRs-MMSNEs-injected tumor under (up) 1 Wcm^{-2} and (middle) 2 Wcm^{-2} irradiations, and (down) PBS solution-injected tumor under 2 Wcm^{-2} irradiation. (d) *In vitro* release profiles of Au NRs-MMSNEs-DOX using dialysis membrane against PBS solution at pH 7.4 and 5.5. (e) A comparison of inhibition rates for MCF-7 cells treated by Au NRs-MMSNEs-NIR (purple), Au NRs-MMSNEs-DOX (red), and Au NRs-MMSNEs-DOX-NIR (green). For photothermal treatment, the media were under 808-nm laser irradiation for 5 min at different power intensities, corresponding to the maximum temperature increases to 39, 42, and 45°C. (Reprinted from Ma M., Chen H., Chen Y., et al., *Biomaterials* 33: 989–998, 2012, with permission from Elsevier Ltd.)

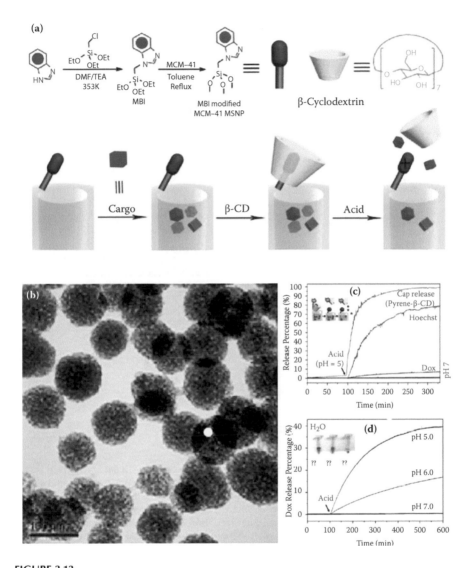

FIGURE 3.13
(a) A graphical representation of the pH-responsive MSNP nanovalve. (b) TEM image of
capped MSNP. (c) Fluorescence intensity plots for the release of Hoechst dye, doxorubicin, and
the pyrene-labeled cyclodextrin cap released from MSNP. (d) Release profiles of doxorubicin
from ammonium-modified (7.5%, w/w) nanoparticles showing the faster and larger response
compared to the unmodified MSNP (c). (Reprinted from Meng H., Xue M., Xia T., et al., *J. Am.
Chem. Soc.* 132: 12690–12697, Copyright 2010, American Chemical Society.)

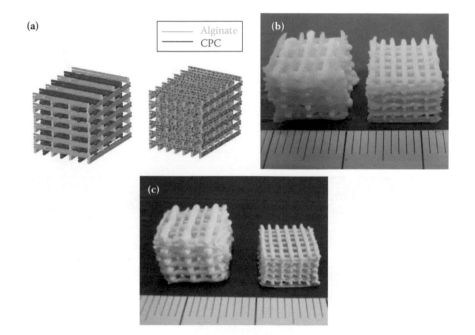

FIGURE 4.10
Biphasic and mixed CPC–alginate scaffolds. (a) CAD models and realized scaffolds in (b) wet and (c) dry state are shown. The biphasic CPC–alginate scaffold (left) was produced by alternate plotting of P-CPC and a high-concentrated alginate–PVA paste through nozzles with an inner diameter of 610 μm (printing speed: 3 mm/s, dosing air pressure: 5.0 bar for the P-CPC and 7.4 bar for the alginate–PVA paste). For fabrication of the mixed CPC–alginate scaffold (right), a CPC–alginate–PVA paste was plotted through a nozzle with inner diameter of 610 μm (printing speed: 3 mm/s, dosing air pressure: 8.5 bar).

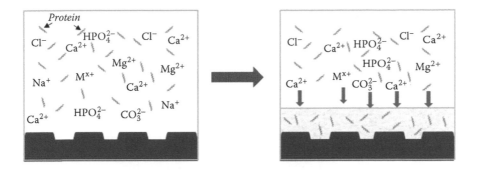

FIGURE 5.2
Schematic process of biomineralization of biological hydroxyapatite in simulated body fluid containing proteins.

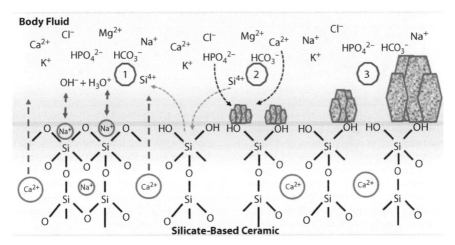

1. Rapid diffusion-controlled ion exchange of network modifying Ca^{2+} and/or Na^+ ions from the ceramic with H_3O^+ ions from the body fluid and silanol groups (Si-OH) at the surface form.
2. HPO_4^{2-} and Ca^{2+} ions from the body fluid are incorporated into the silicate layer from an amorphous calcium phosphate layer (nucleation).
3. Crystallization and growth into biological hydroxyapatite.

FIGURE 5.3
Schematic of the mechanism of formation of calcium phosphate on the surface of silicate based ceramics.

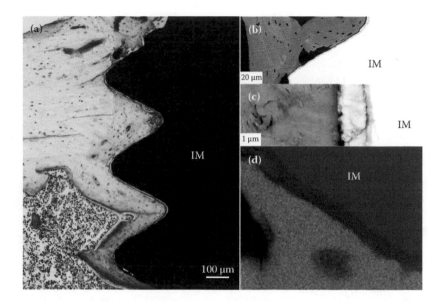

FIGURE 5.5

Sr-HA implants after 28 days of implantation in animal model. (a) Histological sections revealed mineralized bone growth in the medullary area along and in direct contact with surfaces was observed (contact osteogenesis). (b, c) Backscatter scanning electron microscopy micrographs of the implants after 28 days of healing. (b) Low-magnification image showing the implant, implant surface, and bone tissue. Osteocyte lacunae and canaliculi were frequently observed close to the implant surface. (c) Higher-magnification image showing direct contact was observed. (d) Overlapped element maps of calcium (green), titanium (red), and oxygen (blue), showing bone formation along the HAp-implant surface at 28 days healing. The enlarged surface oxide is shown in purple (overlapped blue and red) along the implant perimeter.

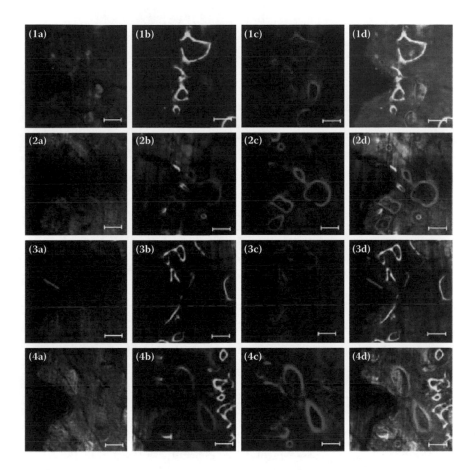

FIGURE 7.2

New bone formation was determined histomorphometrically by bone marker quantification, representing the different healing periods in the four groups, which are enumerated as respective rows in the following matrix of images. Column (a) shows the basic confocal LASER microscope image of samples. Column (b) shows fluorescence of calcein introduced 4 weeks after implantation. Column (c) shows fluorescence of alizarin introduced 8 weeks after implantation. Column (d) shows folded images of the two fluorochromes with the basic confocal image.

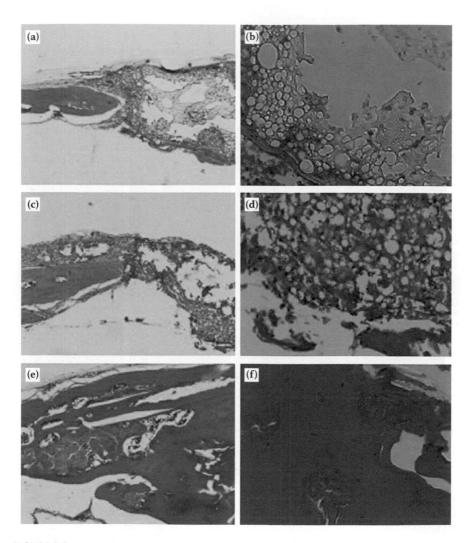

FIGURE 7.8
The *in vivo* bone formation was assessed by (a) H&E for (b) PLGA microspheres, (c, d) for amorphous $CaSiO_3$/PLGA microspheres, and (e, f) for crystal $CaSiO_3$/PLGA microspheres. Panels (a), (c), and (e) are low magnification images by 4 times; (b), (d), and (f) are higher magnification images by 40 times.

T - #0416 - 071024 - C8 - 234/156/11 - PB - 9780367380243 - Gloss Lamination